Springer

Berlin
Heidelberg
New York
Barcelona
Hong Kong
London
Milan
Paris
Singapore
Tokyo

NATALIE EYNARD JUSTIN TEISSIÉ (EDS.)

Electrotransformation of Bacteria

With 14 Figures

Springer

NATALIE EYNARD, PHD
JUSTIN TEISSIÉ, DR., DIRECTEUR DE RECHERCES AU CNRS
Institut de Pharmacologie et Biologie Structurale
CNRS UPR 9062
118 route de Narbonne
31062 Toulouse cédex
France

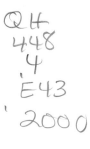

ISBN 3-540-66680-X Springer-Verlag Berlin Heidelberg New York

Library of Congress Cataloging-in-Publication Data
Electrotransformation of bacteria / Natalie Eynard, Justin Teissié (eds.)
  p. cm. – (Springer lab manuals)
Includes bibliographical references and index.
  ISBN 354066680X (lib. bdg. : alk paper)
1. Bacterial transformation – Laboratory manuals.
2. Electric fields. I. Eynard, Natalie, 1964 – II. Teissié Justin, 1947 – III. Series.
  QH448.4.E43 2000
571.9'648-dc21   99-086708

Springer-Verlag is a company in the BertelsmannSpringer publishing group.
© Springer-Verlag Berlin Heidelberg 2000
Printed in Germany

Cover design: design & production GmbH, D-69121 Heidelberg
Typesetting: Mitterweger & Partner, D-68723 Plankstadt
SPIN 10523149   27/3136 5 4 3 2 1 0 – Printed on acid free paper

Preface

Electrotransformation of intact Bacteria was described for the first time only 12 years ago. Since then it has been proved to be a highly efficient technique and easy to use. Gram-positive as well as Gram-negative bacteria can be transformed. Several manufacturers provide reliable electropulsers and in some cases also protocols. But as the molecular mechanisms supporting the introduction of macromolecules into the bacterial cell remain unknown, empirism and adaptation of methods are still the most effective tools to obtain transformed species. Parameters are clearly dependent on the species, which are used. An impressive number of protocols on various species has by now been published and can be used as a data base of protocols for various research fields.

The idea of this lab manual is to provide scientists with protocols for a wide spectrum of species – protocols which had been carefully checked and proved to be effective. They can be used as they are or might be slightly modified to improve the yield. About 40 different species are considered in this manual. Chapters are selected and ordered by fields of applications. The choice can sometimes be rather arbitray since some species could be treated under different application fields. A major part of each chapter is devoted to the specific problems of the species in use and, of course, suggestions of how to find solutions.

Only procedures where intact bacteria are used are included in the manual. Protocols in which protoplasts were first prepared, mixed with plasmids and then electropulsed, are not described. They were observed to be less effective because the survival rate was always very low. This appears to be due to the lethal effect of the field when the plasma membrane is not protected by a wall. Furthermore, these procedures are time consuming. Nevertheless, such an approach may be considered with reluctant species where the integrity of the cell envelope obviously prevents an effective transport.

In the first part of the book, general and technical aspects are treated. Taking into account that the molecular aspects responsible for the induction of electrocompetence of pulsed intact bacteria remain unresolved, we have not tried to describe the different hypothesis currently discussed. One of their limits is that they are often based on the simple assumption that a cell is a sphere built from a lipid bilayer. Very few descriptions take the presence of proteins into account, the ellipsoidal shape and the presence of a wall. How the bacterial envelope is affected remained unknown. In the first chapter we therefore describe our present experimental knowledge on the processes driving plasmids into the cytoplasm when bacteria are pulsed. The second chapter gives a description of the physical, mainly electrical processes, which take place when a field pulse is applied. This should allow the user to estimate the physical limits of an experiment and to avoid the occurence of arcing.

The first protocols are on the laboratory workhorse, E. coli. The, with respect to the importance of bacterial transformation for the biotechnological industry, techniques using strains relevant to Biotechnology and Food Technology are described, followed by protocols for species involved in Medical and Veterinary Applications. The next section deals with the electrotransformation of strains with relevance for the agroindustry. Here, protocols dealing with the modification of Bacteria interacting with plants are given. In the final chapters, procedures on the genetic modifications of bacteria present in the environment, which are now of growing importance, are described in detail. In Appendix A some practical notes on the preparation of plasmids are made, followed by Appendix B, a list of suppliers of the equipment.

We would like to thank all contributors for their works and express our gratitude to many scientists in the CNRS community in Toulouse for their helpful comments and friendly discussions during the last 15 years. This local network was a strong support for the development of our contribution to bacterial electrotransformation.

Toulouse, April 2000   NATALIE EYNARD
                                          AND JUSTIN TEISSIÉ

Contents

Part I Introduction

Chapter 1
General Principles of Bacteria Electrotransformation:
Key Steps
NATALIE EYNARD and JUSTIN TEISSIÉ 1

Chapter 2
A Critical Introduction to the Technology
of Bacterial Electrotransformation
JUSTIN TEISSIÉ 23

Chapter 3
Electrotransformation of *Escherichia coli*
NATALIE EYNARD and JUSTIN TEISSIÉ 35

Chapter 4
Transformation of *Bacillus subtilis* PB1424
by Electroporation
PATRIZIA BRIGIDI, MADDALENA ROSSI
and DIEGO MATTEUZZI 42

Part II Biotechnology and Food Technology

Chapter 5
Clostridium in Biotechnology and Food Technology
HANS P. BLASCHEK, JOSEPH FORMANEK and C.K. CHEN   47

Subprotocol 1: Electroporation-Induced Transformation
              of *Clostridium perfringens* 50
Subprotocol 2: Electroporation Induced Transformation
              of *Clostridium beijerinckii* 53

Chapter 6
Electrotransformation of *Lactococcus lactis*
PASCAL LE BOURGEOIS, PHILIPPE LANGELLA
and PAUL RITZENTHALER 56

Chapter 7
Electrotransformation of *Salmonella typhimurium*
NATALIE EYNARD 66

Chapter 8
Electroporation of bifidobacteria
MADDALENA ROSSI, PATRIZIA BRIGIDI and DIEGO
MATTEUZZI 72

Chapter 9
Electrotransformation of Listeria species
JANET E. ALEXANDER 78

Subprotocol 1: Preparation of Competent Cells for the
                Electroporation of Listeria Species 80
Subprotocol 2: Electroporation of Listeria Species
                with Vectors Greater than 10 kb 85

Chapter 10
Transformation of *Methylobacterium extorquens*
with a Broad-Host-Range Plasmid by Electroporation
SHUNSAKU UEDA and TSUNEO YAMANE 88

Chapter 11
Electrotransformation of Acidophilic, Heterotrophic,
Gram-negative Bacteria
THOMAS E. WARD 94

Chapter 12
Acetobacter xylinum – Biotechnology
and Food Technology
ROBERT E. CANNON 104

Chapter 13
Electrotransformation of *Sphingomonas pancimobilis*
ISABEL SÁ-CORREIA and ARSÉNIO M. FIALHO 108

Chapter 14
Bacillus amyloliquefaciens – Production Host
for Industrial Enzymes
JARI VEHMAANPERÄ 119

Part III Medical and Veterinary Applications

Chapter 15
Electrotransformation of *Yersinia ruckeri*
JUAN M. CUTRÍN, ALICIA E. TORANZO and JUAN L. BARJA 125

Chapter 16
Electrotransformation of Enterococci
ALBRECHT MUSCHOLL-SILBERHORN
and REINHARD WIRTH 134

Chapter 17
Prevotella bryantii, P. ruminicola and *Bacteroides* Strains
HARRY J. FLINT, JENNIFER C. MARTIN
and ANDREW M. THOMSON 140

Chapter 18
Electrotransformation of *Bordetella*
GAVIN R. ZEALEY and REZA K. YACOOB 150

Chapter 19
Transformation of *Campylobacter jejuni*
BEN N. FRY, MARC M.S.M. WÖSTEN,
TRUDY M. WASSENAAR and
BERNARD A.M. VAN DER ZEIJST 157

Chapter 20
Slow-Growing Mycobacteria
BARRY J. WARDS and DESMOND M. COLLINS 168

Chapter 21
Electrotransformation of *Photobacterium damselae*
subsp. *piscicida*
JUAN M. CUTRIN, JUAN L. BARJA and ALICIA E. TORANZO 175

Chapter 22
Actinobacillus actinomycetemcomitans:
Electrotransformation of a Periodontopathogen
KEITH P. MINTZ and PAULA FIVES-TAYLOR 182

Chapter 23
Francisella in Medical and Veterinary Applications
SIOBHÁN C. COWLEY and FRANCIS E. NANO 188

Chapter 24
Electroporation of the Anaerobic Rumen Bacteria
Ruminococcus albus
PIER SANDRO COCCONCELLI 195

Chapter 25
Electroporation of *Legionella* Species
V.K. VISWANATHAN and NICHOLAS P. CIANCIOTTO ... 203

Chapter 26
Electrotransformation of *Streptococcus pneumoniae*
JACQUES LEFRANCOIS and ARMAND MICHEL SICARD .. 212

Part IV Plants

Chapter 27
Clavibacter michiganensis – Transformation of
a Phytopathogenic Gram-Positive Bacterium
DIETMAR MELETZUS, HOLGER JAHR
and RUDOLF EICHENLAUB 219

Chapter 28
Electrotransformation of *Agrobacterium tumefaciens*
and *A. rhizogenes*
DIETHARD MATTANOVICH and FLORIAN RÜKER 227

Part V Environmental Bacteria

Chapter 29
Transformation of the Filamentous Cyanobacterium
Fremyella diplosiphon
ARTHUR R. GROSSMAN and DAVID M. KEHOE 233

Chapter 30
Electroporation of *Bacillus thuringiensis*
and *Bacillus cereus*
JACQUES MAHILLON and DIDIER LERECLUS 242

Chapter 31
Introduction of Plasmids into *Azospirillum brasilense*
by Electroporation
ANN VANDE BROEK and JOS VANDERLEYDEN 253

Chapter 32
Cyanobacteria: Electroporation and Electroextraction
TOIVO KALLAS 257

Chapter 33
Electrotransformation of Plasmids into Freshwater
and Marine Caulobacters
JOHN SMIT, JOHN F. NOMELLINI and WADE H. BINGLE   271

Appendices

Appendix A
Plasmid Preparation
NATALIE EYNARD and JUSTIN TEISSIÉ 283

Appendix B
Suppliers 285

Subject Index 289

Part I

Introduction

General Principles of Bacteria Electrotransformation: Key Steps

NATALIE EYNARD and JUSTIN TEISSIÉ

▧ Introduction

Transfer of foreign information in the genome of cells is a key problem in cell biology and biotechnology. Bioelectrochemistry provided a major innovation when in 1982 Neumann introduced the electrotransformation method (Neumann, et al., 1982). The methodology is simple:

1. cells and plasmids are mixed,

2. an electric field of high intensity is applied,

3. the mixture is incubated to get the expression,

4. a selection assay gives the transformants.

Electrotransformation of intact cells is now routinely used for mammalian and bacterial cells, and more recently demonstrated for yeast (Meilhoc, et al., 1990) and plant cells (Sabri, et al., 1996), protoplasts as well as intact cells. The key problem in transformation is the introduction of DNA into the cell through the envelope. Taking into account the knowledge and hypothesis on the mechanisms of eucaryot's membranes electropermeabilization (Neumann, et al., 1989, Tsong, 1991, Chang, et al., 1992, Zimmermann and Neil, 1996), several models were proposed. But clearly the molecular processes remain unknown in spite of the widespread use of this methodology in cell biology and bio-

Natalie Eynard, IPBS CNRS (UPR 9062), 118 Route de Narbonne, Toulouse, 31062, France

✉ Justin Teissié, IPBS CNRS (UPR 9062), 118 Route de Narbonne, Toulouse, 31062, France (*phone* +33-561-33-58-80; *fax* +33-561-33-58-60; *e-mail* justin@ipbs.fr)

technology. For walled cells, effect of field on the cell wall is not known. But in most cases bacterial wall is a source of reduction of the plasmid entry being a second barrier for the DNA entry into the cell. Earlier works used protoplasts as material for the bacterial electrotransformation (Shivarova, et al., 1983, McIntyre and Harlander, 1989) but then wall regeneration appeared as a non reproducible step which drastically decreased the viability and then the transformation efficiency (Taketo, 1988).

Nevertheless, pulsing the cells has become a convenient way to mediate direct gene transfer into recipient cells (i.e. electrotransformation, (Dower, et al., 1988, Mercenier and Chassy, 1988)). A non exhaustive list of electrotransformed bacterial genera is presented in Table 1. This is a good indication that electropulsation is a general way to cell manipulation.

Table 1. List of electrotransformed bacterial genera constructed according to the Bergey's manual of systematic bacteriology (Krieg and Holt, 1984, Sneath, et al., 1986, Staley, et al., 1989, Williams, et al., 1989)

BERGEY'S SECTION	Genus	
1 Spirochetes	Treponema	(Li and Kuramitsu, 1996)
	Borrelia	(Samuels, 1995)
2 Aerobic/ microaerophilic G⁻	Aquaspirillum	(Eden and Blakemore, 1991)
	Azospirillum	(Broek, et al., 1989)
	Campylobacter	(Miller, et al., 1988)
4 G⁻ Aerobic rods and cocci	Pseudomonas	(Fiedler and Wirth, 1988, Artiguenave, et al., 1997)
	Xanthomonas	(Wang and Tseng, 1992)
	Rhizobium	(Bowen and Kosslak, 1992)
	Bradyrhizobium	(Guerinot, et al., 1990)
	Agrobacterium	(Mattanovich, et al., 1989, McCormac, et al., 1998)
	Legionella	(Cianciotto and Fields, 1992)
	Acinetobacter	(Leahy, et al., 1994)
	Alcaligenes	(Gilis, et al., 1998)
	Brucella	(Tatum, et al., 1993)

Table 1. Continoued

BERGEY'S SECTION	Genus	
	Bordetella	(Zealey, et al., 1988)
	Francisella	(Anthony, et al., 1991)
5 Facultatively anaerobic G- rods	Escherichia	(Dower, et al., 1988, Suzuki, et al., 1997)
	Shigella	(Porter and Dorman, 1997)
	Salmonella	(Sixou, et al., 1991, Toro, et al., 1998)
	Klebsiella	(Trevors, 1990)
	Erwinia	(Ito, et al., 1988)
	Serratia	(Palomar and Vinas, 1996)
	Proteus	(Katenkamp, et al., 1992)
	Yersinia	(Cutrin, et al., 1994)
	Vibrio	(Hamashima, et al., 1990, Cutrin, et al., 1995)
	Pasteurella	(Jablonski, et al., 1992)
	Haemophilus	(Mitchell, et al., 1991)
	Actinobacillus	(Sreenivasan, et al., 1991)
	Zymomonas	(Okamoto and Nakamura, 1992)
6 Anaerobic G⁻ straight, curved and helical rods	Bacteroides	(Thomson and Flint, 1989)
	Butyrivibrio	(Clark, et al., 1994)
7 Dissimilatory sulfate or sulfur reducing bacteria	Desulfovibrio	(Rousset, et al., 1998)
9 Rickettsias and chlamydias	Rochalimea	(Reschke, et al., 1991)
	Bartonella	(Grasseschi and Minnick, 1994)
10 Mycoplasmas	Mycoplasma	(Hedreyda, et al., 1993)
	Spiroplasma	(Foissac, et al., 1997)

Table 1. Continoued

BERGEY'S SECTION	Genus	
12 G⁺ Cocci	Micrococcus	(Grones and Turna, 1995)
	Staphylococcus	(Schenk and Laddaga, 1992)
	Streptococcus	(Suvorov, et al., 1988, Lefrancois, et al., 1998)
	Enterococci	(Fiedler and Wirth, 1988, Yamamoto and Takano, 1996)
	Lactic Acid Streptococci	(Harlander, 1987)
	Leuconostoc	(David, et al., 1989)
	Ruminococcus	(Cocconcelli, et al., 1992)
13 Endospore forming G⁺ rods and cocci	Bacillus	(Belliveau and Trevors, 1989, Mahillon, et al., 1989, Brigidi, et al., 1990)
	Clostridium	(Allen and Blaschek, 1990, Klapatch, et al., 1996)
14 Regular, nonsporing G⁺ rods	Lactobacillus	(Chassy and Flickinger, 1987, Thompson, et al., 1997)
	Listeria	(Alexander, et al., 1990)
15 Irregular, nonsporing G⁺ rods	Coryne-bacterium	(Haynes and Britz, 1989, Ankri, et al., 1996)
	Brevibacterium	(Haynes and Britz, 1989, Satoh, et al., 1990, Chan Kwo Chion, et al., 1991)
	Actinomyces	(Yeung and Kozelsky, 1994, Jost, et al., 1997)
	Bifidobacterium	(Missich, et al., 1994, Matsumura, et al., 1997)
16 Mycobacteria	Mycobacterium	(Hermans, et al., 1990, Wards and Collins, 1996)
17 Nocardioforms	Nocardia	(Valentin and Dennis, 1996)
	Rhodococcus	(Desomer, et al., 1990, Sunairi, et al., 1996)

Table 1. Continoued

BERGEY'S SECTION	Genus	
18 Anoxygenic phototrophic bacteria	Purple non sulfur bacteria	(Donohue and Kaplan, 1991)
19 Oxygenic photosynthetic bacteria	Gloeobacter	(Muhlenhoff and Chauvat, 1996)
	Anabaena	(Thiel and Poo, 1989)
	Nostoc	(Moser, et al., 1995)
20 Aerobic chemolithotrophic bacteria and associated organisms	Nitrosomonas	(Hommes, et al., 1996)
	Thiobacillus	(Kusano, et al., 1992)
	Acidiphilium	(Glenn, et al., 1992)
21 Budding and/ or appengaged bacteria	Caulobacter	(Gilchrist and Smit, 1991)
23 Non photosynthetic, non fruiting, gliding bacteria	Lysobacter	(Lin and McBride, 1996)
24 fruiting, gliding acteria	Myxococcus	(Ramaswamy, et al., 1997)
25 Archaeobacteria	Methanococcus	(Micheletti, et al., 1991)
27 Actinomyces with multi locular sporangia	Frankia	(Cournoyer and Normand, 1992)
29 Streptomyces	Streptomyces	(Macneil, 1987, Mazy-Servais, et al., 1997)

No molecular description of the events affecting the envelope, the inner membrane as well as plasmid can be given at the present state of our knowledge. A phenomenological description is that the field is able to induce an inner membrane permeabilization (shown by the resulting ATP leakage). This appears to be linked to the induction of an overcritical membrane potential during the pulse (several milliseconds). This permeabilization remains present for seconds or minutes after the pulse.

The efficiency of electrotransformation of *E. coli* is strongly correlated to the level of electropermeabilization (Sixou, et al., 1991), although mechanisms were different (Eynard, et al., 1992). A systematic investigation of *E. coli* electrotransformation was described in a serie of 3 papers by Tsong's group (Xie, et al., 1990, Xie and Tsong, 1990, Xie and Tsong, 1992). The DNA was proposed to be first absorbed on the cell surface but able to diffuse on the cell surface. Electropulsation then induced electropores where a free transfer of absorbed plasmids would take place. In such a case, transmembrane transfer was a postpulse event but the DNA absorption could not take place after the pulse.

DNA transfer was not due just to holes punched in the lipid matrix. In processes associated with electro-permeabilisation such as electrotransformation, an active role of the cell membrane was clearly evident.

1 Pre Pulse Events

Preparation of "electrocompetent" cells

Results show that each treatment which increases the plasmid interfacial concentration leads to an enhanced transformation rate. The cell envelope quality (even for species easily electrotransformable) influences the transformation efficiency but mechanisms are unknown.

Effect of the culture medium composition and the growth temperature

For some species the nature of the growth medium or the growth temperature appears to have an effect on the transformation ef-

ficiency. For example, an improved efficiency was obtained on
S. aureus by the use of a yeast extract containing medium rather
than the conventional SOC medium (Schenk and Laddaga,
1992), and the same effect was observed for *E. coli* C (Taketo,
1989). The effect of the growth temperature has been studied
but results are not apparent in the same way, for *E. coli* XL-1
Blue an increase of the transformation efficiency was obtained
by growth at 18° instead of the classical 37° (Chuang, et al., 1995),
but the same effect was obtained by using a growth temperature
of 44°C for *E. coli* C. In fact it seems that in all studies the classical
growth temperature (in terms of cellular growth rate) is not the
optimal growth temperature for preparation of electrocompetent cells (Glenn, et al., 1992).

Chemical or enzymatic alterations of the wall

The step of plasmid absorbtion on the cell membrane could be
slowed or blocked by the cell wall presence. The experimental
strategy used is the loss of the wall organization by biochemical
or physical methods during, or after the growth, but always
before the pulse. In bacterial electrotransformation, chemical
treatments derived from lysing or protoplasting protocols
were applied. These treatments affect a metabolic pathway
or a specific component of the cell wall and are highly strain-
specific. Some of the most common treatments are presented:

Glycine (Dunny, et al., 1991, Shepard and Gilmore, 1995, Sunairi, et al., 1996) is used to increase the plasmid transfer through
the cell wall of some Gram$^+$ bacteria or Nocardioforms. Glycine
can replace D-alanine of the peptidoglycan precursors and,
being a poor substrate for the transpeptidation reaction, the
cross linking of the peptidoglycan is then reduced and the
cell wall is more fragile.

A racemic mixture of D and L-threonine enhanced the electrotransformation of *Lactococcus lactis* (McIntyre and Harlander, 1989) or some strains of *Bacillus subtilis* not by a direct incorporation in the peptidoglycan structure but by an inhibition
of the diaminopimelic acid (DAP) incorporation in peptidoglycan (McDonald, et al., 1995).

Mild treatment of some gram$^+$ bacteria with lysozyme increased the transformation efficiency for Lactococcus (Powell,

et al., 1988), by partially removing the cell wall (Strominger and Ghuysen, 1967), the glycocalyx, or both.

A treatment specific by Isonicotinic acid hydrazide (INH) as inhibitor of mycolic acid synthesis is used for mycobacteria (Hermans, et al., 1990) and other mycolic-acid-containing bacteria as *Corynebacterium glutamicum* (Haynes and Britz, 1990). For these species, Tween 80, which is known to prevent clumping during growth of mycobacteria and to change the mycolic acid composition of cutaneous corynebacteria, could enhance the electrotransformation efficiency when added during growth (Haynes and Britz, 1989).

Storage of cells

As a general rule, electrocompetent cells can be prepared and stored. Cells frozen in glycerol or DMSO can be pulsed after thawing without any decrease of the transformation efficiency (Diver, et al., 1990, Schenk and Laddaga, 1992, Chuang, et al., 1995).

Effects of size and molecular form of DNA and plasmid concentration

The physical properties of the plasmid used in experiments is an important factor in electrotransformation. The open question in the field of cell transformation by entry of a plasmid is that the transformation of the cell is synergy of a correct entry of the plasmid and a high level of its expression. Experiments on effect of size or molecular form of plasmids have not yet discriminated these two steps. Effect is observed in the total response in terms of number of cells electrotransformed. Many experiments appear to show that the transformation efficiency decreased with an increase in the size of DNA, but a limit to this conclusion is that variation of the size is obtained by use of different plasmids with various antibiotic markers and origins (McLaughlin and Ferretti, 1995, Szostkova and Horakova, 1998). This observation is not obvious when different plasmid sizes were obtained either by ligation of the same marker region in various plasmids or by successive deletion in a given plasmid (Allen and Blaschek, 1990, Ohse, et al., 1995). In these conditions, under 25 kb, plasmid size does not appear to have an effect on the transformation efficiency. Nevertheless, for larger DNA molecules (more than 50-100kb), the decrease of transformation efficiency is more evident (Sheng, et al., 1995).

Transformation efficiency is markedly affected by molecular form of DNA used for the electrotransformation. Cell electrotransformation with circular supercoiled DNA gives the maximum efficiency, higher than for the circular relaxed DNA (Ohse, et al., 1997). The transformation efficiency for the linearized DNA is extremely low or non-existant (Xie, et al., 1992, Kimoto and Taketo, 1996, Ohse, et al., 1997).

Time of plasmid addition

It seems that an increase of transformation efficiency is related to a better contact between cells and plasmid. Performing the pre incubation at low temperature could limit extracellular nuclease activity and then reduce degradation of plasmid DNA.

The time addition of the plasmids to bacteria (Eynard, et al., 1992) plays a critical role in yield of transformants for each cell type. For all species tested, when plasmid was added before the pulse, yield of transformants remained constant or slightly decreased when the prepulse incubation duration is increased. But if DNA was added even a few seconds after the pulse, no transformed cells was observed.

2 Interaction of Plasmid and Cell Envelope During Electropulsation

Correlation between transformation efficiency and permeabilized cell surface

Electropermeabilisation to small molecules is triggered as soon as the field induced membrane potential difference reaches a critical value (200-250 mV) (Teissié and Rols, 1993). Electrotransformation is observed only when cells are pulsed with field strengths larger than that which is needed to induce permeabilisation. The extent of permeabilisation is then controlled by the field strength. Gene transfer should occur in the part of the cell surface which is brought to the state where free diffusion of small molecules can take place.

One limit is that transformation is controlled by the cell viability. When drastic field intensities are used, the loss in viability (Sixou, et al., 1991) is associated by a drop in transformability of surviving cells.

Orientation process. Consequence on pulse duration effect

The effect of electric fields on cells is not limited to the induction of a membrane potential difference (Zimmermann, 1982). Among other consequences, the Joule heating is always limited by the use of a low ionic content buffer (i.e. a low current). But one should take into account the induction of dipoles on the cell surface and the orientation process linked to the torque created by the interaction of the field on the endogenous and induced cell dipoles (O'Konski and Haltner, 1957). This is very important in the case of rod shaped bacteria.

For a non spherical cell, the first event during the pulse is the rod orientation with the long axis parallel to the field lines (Eynard, et al., 1992, Kimoto and Taketo, 1997). Orientation process is fast (characteristic time 1 ms for E. coli), electropermeabilization occurs mostly when the orientation phase is finished (Eynard, et al., 1998). The presence of this orientation step explains the poor effect of short pulse duration (shorter than 1 ms) in the case of bacteria even if strong fields are applied. Long pulse duration, even when associated to a lower field amplitude (but always higher than the permeabilization threshold) will be more efficient in the case of rod shaped bacteria.

DNA – membrane have a strong interaction during the pulse

When electrooptics experiments are carried out in presence of plasmid under electrical conditions prone to induce electro-transformation, results show that presence of plasmid affects orientation and permeabilization kinetics (Eynard, et al., 1998).

Presence of DNA in the cell suspension during the pulse slows down the orientation of the rod. The orientational time for *E. coli* is slowed (about two-fold) when DNA is present. Permeabilization kinetic is on the contrary accelerated (about ten-fold) even if steady state of permeabilization remains the same (i.e: extent of ATP leakage is the same in presence or in absence of plasmid). The field dependent processes are very complex. Besides the orientation of the rod, they reflect an alteration of the outer membrane and of the inner membrane.

It appears that DNA and bacterial envelope have a strong interaction during the pulse. The molecular nature of this process is not known but as permeabilization kinetic (and then transient structures of permeabilization) were modified, it seems that interaction could be assimilated to an "envelope anchorage" of the plasmid.

3 Post Pulse Events

DNAase I effect

Post pulse DNAase treatment of the plasmid/cells mixture affects the yield of transformation. For bacteria (Eynard, et al., 1992, Kimoto and Taketo, 1997) addition of DNAase I just after the pulse provoked a decrease in transformation efficiency. Duration of DNAase effect is short for bacteria (less than 10 seconds). But the sensitivity to DNAase is present for a duration which is more longer than the electric pulse. DNAase I doesn't affect cell viability and has no effect on the resistance of a pulsed bacteria containing a plasmid.

DNA remains able to be cleaved by external DNAase some seconds after the pulse. This indicates that DNA entry in the cytoplasm was not completely achieved during the pulse. DNA entry is not only dependent on a direct field effect. It is a much slower process (3 orders of magnitude) than the ones which occur during the pulse. Effect of physiological parameters just after the pulse could then control the transformation efficiency.

Effect of post-pulse temperature

Conflicting data on the temperature dependence of electrotransformation have been reported. A free diffusion of DNA across field created electropores should occur to a larger extent if the lifetime of the permeabilized state was very long. As life time of permeabilization depends on temperature, the common conditions described in literature are incubation at 4°C after the pulse (Potter, 1993). The temperature dependence of electrotransformation was controlled (Eynard et al., 1992). A tenfold increase in transformation yield is obtained if pulsed cells were diluted in warm medium (37°C) just after pulsing rather than kept on ice for 10 minutes before dilution. This observation supports an active cell participation in the DNA entry after the pulse.

4 Expression of Results

Two classical expressions of the transformation are used (Szostkova and Horakova, 1998). The efficiency of transformation is reflected by the number of transformants / μg of used DNA and the transformation frequency is calculated by the percentage of survival cells which have been transformed. The transformation efficiency is related to the initial cell number used in the experiments, then the result includes the cell mortality and the transformation while transformation frequency reports only the transformability of the survival cells. To compare results obtained with various protocols, apparatus or bacterial strain, the transformation efficiency is more adapted. The only limit to the use of this value is the absence in some protocols of a detailed "materials and methods" part. This does not allow us to know the exact number or concentration of pulsed bacteria. Yet the efficiency of transformation depends on the treated cell concentration. Another expression of the transformation efficiency is the number of transformants / μg of used DNA / number of treated cells.

When different DNA sizes are used, the transformation efficiency is misleading because the number of DNA molecules / μg vary with DNA molecular weight and transformation efficiency is related to the DNA molecules number. The best way to compare various DNA for transformation is to express results in terms of molecular efficiency as transformants per molecule of input DNA (Ohse, et al., 1995).

References

Alexander, J. E., P. W. Andrew, D. Jones and I. S. Roberts. 1990. Development of an optimized system for electroporation of Listeria species. Lett Appl Microbiol. 10:179-181

Allen, S. P. and H. P. Blaschek. 1990. Factors involved in the electroporation-induced transformation of Clostridium perfringens. Fems Microbiol Lett. 58:217-220

Ankri, S., O. Reyes and G. Leblon. 1996. Improved electro-transformation of highly DNA-restrictive corynebacteria with DNA extracted from starved Escherichia coli. Fems Microbiol Lett. 140:247-251

Anthony, L. S., M. Z. Gu, S. C. Cowley, W. W. Leung and F. E. Nano. 1991. Transformation and allelic replacement in Francisella spp. Journal of General Microbiology. 137:2697-2703

Artiguenave, F., R. Vilagines and C. Danglot. 1997. High-efficiency transposon mutagenesis by electroporation of a Pseudomonas fluorescens strain. Fems Microbiol Lett. 153:363-369

Belliveau, B. H. and J. T. Trevors. 1989. Transformation of Bacillus cereus vegetative cells by electroporation. Appl Environ Microbiol. 55:1649-1652

Bowen, B. A. and R. M. Kosslak. 1992. Electrical energy changes conductivity and determines optimal electrotransformation frequency in Gram-negative bacteria. Appl Environ Microbiol. 58:3292-3296

Brigidi, P., E. De Rossi, M. L. Bertarini, G. Riccardi and D. Matteuzzi. 1990. Genetic transformation of intact cells of Bacillus subtilis by electroporation. Fems Microbiol Lett. 55:135-138

Broek, A. V., A. Van Gool and J. Vanderleyden. 1989. Electroporation of Azospirillum brasilense with plasmid DNA. Fems Microbiol Lett. 61:177-182

Chan Kwo Chion, C. K. N., R. Duran, A. Arnaud and P. Galzy. 1991. Electrotransformation of whole cells of Brevibacterium sp. R312 a nitrile hydratase producing strain: construction of a cloning vector. Fems Microbiol Lett. 81:177-184

Chang, D. C., B. M. Chassy, J. A. Saunders and A. E. Sowers. 1992. Guide to electroporation and electrofusion in Academic press, London.

Chassy, B. M. and J. L. Flickinger. 1987. Transformation of Lactobacillus casei by electroporation. Fems Microbiol Lett. 44:173-177

Chuang, S. E., A. L. Chen and C. C. Chao. 1995. Growth of E. coli at low temperature dramatically increases the transformation frequency by electroporation. Nucleic Acids Res. 23:1641

Cianciotto, N. P. and B. S. Fields. 1992. Legionella pneumophila mip gene potentiates intracellular infection of protozoa and human macrophages. Proc Natl Acad Sci. 89:5188-5191

Clark, R. G., K. J. Cheng, L. B. Selinger and M. F. Hynes. 1994. A conjugative transfer system for the rumen bacterium, Butyrivibrio fibrisolvens, based on Tn916-mediated transfer of the Staphylococcus aureus plasmid pUB110. Plasmid. 32:295-305

Cocconcelli, P. S., E. Ferrari, F. Rossi and V. Bottazzi. 1992. Plasmid transformation of Ruminococcus albus by means of high-voltage electroporation. Fems Microbiology Letters. 73:203-207

Cournoyer, B. and P. Normand. 1992. Relationship between electroporation conditions, electropermeability and respiratory activity for Frankia strain ACN14a. Fems Microbiol Lett. 73:95-99

Cutrin, J. M., R. F. Conchas, J. L. Barja and A. E. Toranzo. 1994. Electrotransformation of Yersinia ruckeri by plasmid DNA. Microbiologia. 10:69-82

Cutrin, J. M., A. E. Toranzo and J. L. Barja. 1995. Genetic transformation of Vibrio anguillarum and Pasteurella piscicida by electroporation. Fems Microbiol Lett. 128:75-80

David, S., G. Simons and W. M. De Vos. 1989. Plasmid transformation by electroporation of Leuconostoc paramesenteroides and its use in molecular cloning. Appl Environ Microbiol. 55:1483-1489

Desomer, J., P. Dhaese and M. Van Montagu. 1990. Transformation of Rhodococcus fascians by high voltage electroporation and development of R. fascians cloning vectors. Appl Env Microbiol. 2818-2825

Diver, J. M., L. E. Bryan and P. A. Sokol. 1990. Transformation of Pseudomonas aeruginosa by electroporation. Anal Biochem. 189:75-79

Donohue, T. J. and S. Kaplan. 1991. Genetic techniques in rhodospirillaceae. Methods Enzymol. 204:459-485

Dower, W. J., J. F. Miller and C. W. Ragsdale. 1988. High efficiency transformation of E. coli by high voltage electroporation. Nucleic Acids Res. 16:6127-6145

Dunny, G. M., L. N. Lee and D. J. LeBlanc. 1991. Improved electroporation and cloning vector system for gram-positive bacteria. Appl Environ Microbiol. 57:1194-1201

Eden, P. A. and R. P. Blakemore. 1991. Electroporation and conjugal plasmid transfer to members of the genus Aquaspirillum. Arch Microbiol. 155:449-452

Eynard, N., F. Rodriguez, J. Trotard and J. Teissie. 1998. Electrooptics studies of Escherichia coli electropulsation: orientation, permeabilization, and gene transfer. Biophys J. 75:2587-2596

Eynard, N., S. Sixou, N. Duran and J. Teissie. 1992. Fast kinetics studies of Escherichia coli electrotransformation. Eur J Biochem. 209:431-436

Fiedler, S. and R. Wirth. 1988. Transformation of bacteria with plasmid DNA by electroporation. Anal Biochem. 170:38-44

Foissac, X., C. Saillard and J. M. Bove. 1997. Random insertion of transposon Tn4001 in the genome of Spiroplasma citri strain GII3. Plasmid. 37:80-86

Gilchrist, A. and J. Smit. 1991. Transformation of freshwater and marine caulobacters by electroporation. J Bacteriol. 173:921-925

Gilis, A., P. Corbisier, W. Baeyens, S. Taghavi, M. Mergeay and D. van der Lelie. 1998. Effect of the siderophore alcaligin E on the bioavailability of Cd to Alcaligenes eutrophus CH34. J Ind Microbiol Biotechnol. 20:61-68

Glenn, A. W., F. F. Roberto and T. E. Ward. 1992. Transformation of Acidiphilium by electroporation and conjugation. Can J Microbiol. 38:387-393

Grasseschi, H. A. and M. F. Minnick. 1994. Transformation of Bartonella bacilliformis by electroporation. Can J Microbiol. 40:782-786

Grones, J. and J. Turna. 1995. Transformation of microorganisms with the plasmid vector with the replicon from pAC1 from Acetobacter pasteurianus. Biochem Biophys Res Commun. 206:942-947

Guerinot, M. L., B. A. Morisseau and T. Klapatch. 1990. Electroporation of Bradyrhizobium japonicum. Mol Gen Genet. 221:287-290

Hamashima, H., T. Nakano, S. Tamura and T. Arai. 1990. Genetic transformation of Vibrio parahaemolyticus, Vibrio alginolyticus, and Vibrio cholerae non O-1 with plasmid DNA by electroporation. Microbiol Immunol. 34:703-708

Harlander, S. K. 1987. Transformation of Streptococcus lactis by electroporation in Streptococcal genetics ASM, Washington. 229-233

Haynes, J. A. and M. G. Britz. 1989. Electrotransformation of Brevibacterium lactofermentum and corynebacterium glutamicum: growth in tween 80 increases transformation frequencies. Fems Microbiol Lett. 61:329-334

Haynes, J. A. and M. G. Britz. 1990. The effect of growth conditions of Corynebacterium glutamicum on the transformation frequency obtained by electroporation. J Gen Microbiol. 136:255-263

Hedreyda, C. T., K. K. Lee and D. C. Krause. 1993. Transformation of Mycoplasma pneumoniae with Tn4001 by electroporation. Plasmid. 30:170-175

Hermans, J., J. G. Boschloo and J. A. M. De Bont. 1990. Transformation de Mycobacterium aurum by electroporation: the use of glycine, lysozyme and isonicotinic acid hydrazide in enhancing transformation efficiency. Fems Microbiol Lett. 72:221-224

Hommes, N. G., L. A. Sayavedra-Soto and D. J. Arp. 1996. Mutagenesis of hydroxylamine oxidoreductase in Nitrosomonas europaea by transformation and recombination. J Bacteriol. 178:3710-3714

Ito, K., T. Nishida and K. Izaki. 1988. Application of electroporation for transformation in Erwinia carotovora. Agric Biol Chem. 52:293-294

Jablonski, L., N. Sriranganathan, S. M. Boyle and G. R. Carter. 1992. Conditions for transformation of Pasteurella multocida by electroporation. Microb Pathog. 12:63-68

Jost, B. H., S. J. Billington and J. G. Songer. 1997. Electroporation-mediated transformation of Arcanobacterium (Actinomyces) pyogenes. Plasmid. 38:135-140

Katenkamp, U., I. Groth, F. Laplace and H. Malke. 1992. Electrotransformation of the stable L-form of proteus mirabilis". Fems Microbiol Lett. 94:19-22

Kimoto, H. and A. Taketo. 1996. Studies on electrotransfer of DNA into Escherichia coli: effect of molecular form of DNA. Biochim. Biophys. Acta, Gene Struct. Expr. 1307:325-330

Kimoto, H. and A. Taketo. 1997. Initial stage of DNA-electrotransfer into E. coli cells. J Biochem. 122:237-242

Klapatch, T. R., M. L. Guerinot and L. R. Lynd. 1996. Electrotransformation of Clostridium thermosaccharolyticum. J Ind Microbiol. 16:342-347

Krieg, N. R. and J. G. Holt. 1984. Bergey's manual of systematic bacteriology in Williams & Wilkins, Baltimore.

Kusano, T., K. Sugawara, C. Inoue, T. Takeshima, M. Numata and T. Shiratori. 1992. Electrotransformation of Thiobacillus ferrooxidans with plasmids containing a mer determinant. J Bacteriol. 174:6617-6623

Leahy, J. G., J. M. Jones-Meehan and R. R. Colwell. 1994. Transformation of acinetobacter calcoaceticus RAG-1 by electroporation. Can J Microbiol. 40:233-236

Lefrancois, J., M. M. Samrakandi and A. M. Sicard. 1998. Electrotransformation and natural transformation of Streptococcus pneumoniae: requirement of DNA processing for recombination [In Process Citation]. Microbiology. 144:3061-3068

Li, H. and H. K. Kuramitsu. 1996. Development of a gene transfer system in Treponema denticola by electroporation. Oral Microbiol Immunol. 11:161-165

Lin, D. and M. J. McBride. 1996. Development of techniques for the genetic manipulation of the gliding bacteria Lysobacter enzymogenes and Lysobacter brunescens. Can J Microbiol. 42:896-902

Macneil, D. J. 1987. Introduction of plasmid DNA into Streptomyces lividans by electroporation. Fems Microbiol Lett. 42:239-244

Mahillon, J., W. Chungjapornchai, J. Decock, S. Dierickx, F. Michiels, M. Peferoen and H. Joos. 1989. Transformation of Bacillus thuringiensis by electroporation. Fems Microbiol Lett. 60:205-210

Matsumura, H., A. Takeuchi and Y. Kano. 1997. Construction of Escherichia coli-Bifidobacterium longum shuttle vector transforming B. longum 105-A and 108-A. Biosci Biotechnol Biochem. 61:1211-1212

Mattanovich, D., F. Ruker, A. C. Machado, M. Laimer, F. Regner, H. Steinkellner, G. Himmler and H. Katinger. 1989. Efficient transformation of Agrobacterium spp. by electroporation. Nucleic Acids Res. 17:6747

Mazy-Servais, C., D. Baczkowski and J. Dusart. 1997. Electroporation of intact cells of Streptomyces parvulus and Streptomyces vinaceus. Fems Microbiol Lett. 151:135-138

McCormac, A. C., M. C. Elliott and D. F. Chen. 1998. A simple method for the production of highly competent cells of Agrobacterium for transformation via electroporation. Mol Biotechnol. 9:155-159

McDonald, I. R., P. W. Riley, R. J. Sharp and A. J. McCarthy. 1995. Factors affecting the electroporation of Bacillus subtilis. J Appl Bacteriol. 79:213-218

McIntyre, D. A. and S. K. Harlander. 1989. Genetic transformation of intact Lactococcus lactis subsp. lactis by high-voltage electroporation. Appl Environ Microbiol. 55:604-610

McIntyre, D. A. and S. K. Harlander. 1989. Improved electroporation efficiency of intact Lactococcus lactis subsp. lactis cells grown in defined media. Appl Environ Microbiol. 55:2621-2626

McLaughlin, R. E. and J. J. Ferretti. 1995. Electrotransformation of Streptococci. Methods Mol Biol. 47:185-193

Meilhoc, E., J. M. Masson and J. Teissie. 1990. High efficiency transformation of intact yeast cells by electric field pulses. Biotechnology. 8:223-227

Mercenier, A. and B. M. Chassy. 1988. Strategies for the development of bacterial transformation systems. Biochimie. 70:503-517

Micheletti, P. A., K. A. Sment and J. Konisky. 1991. Isolation of a coenzyme M-auxotrophic mutant and transformation by electroporation in Methanococcus voltae. J Bacteriol. 173:3414-3418

Miller, J. F., W. J. Dower and L. S. Tompkins. 1988. High-voltage electroporation of bacteria: genetic transformation of Campylobacter jejuni with plasmid DNA. Proc Natl Acad Sci U S A. 85:856-860

Missich, R., B. Sgorbati and D. J. LeBlanc. 1994. Transformation of Bifidobacterium longum with pRM2, a constructed Escherichia coli-B. longum shuttle vector. Plasmid. 32:208-211

Mitchell, M. A., K. Skowronek, L. Kauc and S. H. Goodgal. 1991. Electroporation of Haemophilus influenzae is effective for transformation of plasmid but not chromosomal DNA. Nucleic Acids Res. 19:3625-3628

Moser, D., D. Zarka, C. Hedman and T. Kallas. 1995. Plasmid and chromosomal DNA recovery by electroextraction of cyanobacteria. Fems Microbiol Lett. 128:307-313

Muhlenhoff, U. and F. Chauvat. 1996. Gene transfer and manipulation in the thermophilic cyanobacterium Synechococcus elongatus. Mol Gen Genet. 252:93-100

Neumann, E., M. Schaefer-Ridder, Y. Wang and P. H. Hofschneider. 1982. Gene transfer into mouse lyoma cells by electroporation in high electric fields. EMBO J. 1:841-845

Neumann, E., A. E. Sowers and C. A. Jordan. 1989. Electroporation and electrofusion in cell biology in Plenum press, New York, London.

O'Konski, C. T. and A. J. Haltner. 1957. Electric properties of macromolecules. I. A study of electric polarization in polyelectrolyte solutions by means of electric birefringence. J. Am. Chem. Soc. 79:5634- 5648

Ohse, M., K. Kawade and H. Kusaoke. 1997. Effects of DNA topology on transformation efficiency of Bacillus subtilis ISW1214 by electroporation. Biosci Biotechnol Biochem. 61:1019-1021

Ohse, M., K. Takahashi, Y. Kadowaki and H. Kusaoke. 1995. Effects of plasmid DNA sizes and several other factors on transformation of Bacillus subtilis ISW1214 with plasmid DNA by electroporation. Biosci Biotechnol Biochem. 59:1433-1437

Okamoto, T. and K. Nakamura. 1992. Simple and highly efficient transformation method for Zymomonas mobilis: electroporation. Biosci Biotech Biochem. 56:833

Palomar, J. and M. Vinas. 1996. The effect of O-antigen on transformation efficiency in Serratia marcescens. Microbiologia. 12:435-438

Porter, M. E. and C. J. Dorman. 1997. Virulence gene deletion frequency is increased in Shigella flexneri following conjugation, transduction, and transformation. Fems Microbiol Lett. 147:163-172

Potter, H. 1993. Application of electroporation in recombinant DNA technology. Methods Enzymol. 217:461-483

Powell, I. B., M. G. Achen, A. J. Hillier and B. E. Davidson. 1988. A simple and rapid method for genetic transformation of lactic Streptococci by electroporation. Appl Environ Microbiol. 54:655-660

Ramaswamy, S., M. Dworkin and J. Downard. 1997. Identification and characterization of Myxococcus xanthus mutants deficient in calcofluor white binding. J Bacteriol. 179:2863-2871

Reschke, D. K., M. E. Frazier and L. P. Mallavia. 1991. Transformation and genomic restriction mapping of Rochalimaea spp. Acta Virol. 35:519-525

Rousset, M., L. Casalot, B. J. Rapp-Giles, Z. Dermoun, P. de Philip, J. P. Belaich and J. D. Wall. 1998. New shuttle vectors for the introduction of cloned DNA in Desulfovibrio. Plasmid. 39:114-122

Sabri, N., B. Pelissier and J. Teissié. 1996. Transient and stable electrotransformations of intact black Mexican sweet maize cells are obtained after preplasmolysis. Plant Cell Report. 15:924-928

Samuels, D. S. 1995. Electrotransformation of the spirochete Borrelia burgdorferi. Methods Mol Biol. 47:253-259

Satoh, Y., K. Hatakeyama, K. Kohama, M. Kobayashi, Y. Kurusu and H. Yukawa. 1990. Electrotransformation of intact cells of Brevibacterium flavum MJ-233. J Ind Microbiol. 5:159-165

Schenk, S. and R. A. Laddaga. 1992. Improved method for electroporation of Staphylococcus aureus. Fems Microbiol Lett. 73:133-138

Sheng, Y., V. Mancino and B. Birren. 1995. Transformation of Escherichia coli with large DNA molecules by electroporation. Nucleic Acids Res. 23:1990-1996

Shepard, B. D. and M. S. Gilmore. 1995. Electroporation and efficient transformation of Enterococcus faecalis grown in high concentrations of glycine. Methods Mol Biol. 47:217-226

Shivarova, N., W. Förster, H. E. Jacob and R. Grigorova. 1983. Microbiological implications of electric field effects. VII. Stimulation of plasmid transformation of Bacillus cereus protoplasts by electric field pulses. Z All Mikrobiol. 23:595-599

Sixou, S., N. Eynard, J. M. Escoubas, E. Werner and J. Teissie. 1991. Optimized conditions for electrotransformation of bacteria are related to the extent of electropermeabilization. Biochim Biophys Acta. 1088:135-138

Sneath, P. H., N. S. Mair, M. E. Sharpe and J. G. Holt. 1986. Bergey's manual of systematic bacteriology in Williams & Wilkins, Baltimore.

Sreenivasan, P. K., D. J. LeBlanc, L. N. Lee and P. Fives-Taylor. 1991. Transformation of Actinobacillus actinomycetemcomitans by electroporation, utilizing constructed shuttle plasmids [published erratum appears in Infect Immun 1992 Apr;60(4):1728]. Infect Immun. 59:4621-4627

Staley, J. T., M. P. Bryant, N. Pfennig and J. G. Holt. 1989. Bergey's manual of systematic bacteriology in Williams & Wilkins, Baltimore.

Strominger, J. L. and J. M. Ghuysen. 1967. Mechanisms of enzymatic bacteriolysis. Cell walls of bacteria are solubilized by action of either specific carbohydrases or specific peptidases. Science. 156:213-221

Sunairi, M., N. Iwabuchi, K. Murakami, F. Watanabe, Y. Ogawa, H. Murooka and M. Nakajima. 1996. Effect of penicillin G on the electroporation of Rhodococcus rhodochrous CF222. Lett Appl Microbiol. 22:66-69

Suvorov, A., J. Kok and G. Venema. 1988. Transformation of group A streptococci by electroporation. Fems Microbiol Lett. 56:95-100

Suzuki, K., N. Wakao, Y. Sakurai, T. Kimura, K. Sakka and K. Ohmiya. 1997. Transformation of Escherichia coli with a large plasmid of Acidiphilium multivorum AIU 301 encoding arsenic resistance. Appl Environ Microbiol. 63:2089-2091

Szostkova, M. and D. Horakova. 1998. The effect of plasmid DNA sizes and the other factors on electrotransformation of Escherichia coli JM109. Bioelectrochem. Bioenerg. 47:319-323

Taketo, A. 1988. DNA transfection of Escherichia coli by electroporation. Biochim Biophys Acta. 949:318-324

Taketo, A. 1989. Properties of electroporation-mediated DNA transfer in Escherichia coli. J Biochem. 105:813-817

Tatum, F. M., D. C. Morfitt and S. M. Halling. 1993. Construction of a Brucella abortus RecA mutant and its survival in mice. Microb Pathog. 14:177-185

Teissié, J. and M. P. Rols. 1993. An experimental evaluation of the critical potential difference inducing cell membrane electropermeabilization. Biophys. J. 65:409-413

Thiel, T. and H. Poo. 1989. Transformation of a filamentous cyanobacterium by electroporation. J Bacteriol. 171:5743-5746

Thompson, J. K., K. J. McConville, C. McReynolds and M. A. Collins. 1997. Electrotransformation of Lactobacillus plantarum using linearized plasmid DNA. Lett Appl Microbiol. 25:419-425

Thomson, A. M. and H. J. Flint. 1989. Electroporation induced transformation of Bacteroides ruminicola and Bacteroides uniformis by plasmid DNA. Fems Microbiol Lett. 52:101-104

Toro, C. S., G. C. Mora and N. Figueroa-Bossi. 1998. Gene transfer between related bacteria by electrotransformation: mapping Salmonella typhi genes in Salmonella typhimurium. J Bacteriol. 180:4750-4752

Trevors, J. T. 1990. Electroporation and expression of plasmid pBR322 in Klebsiella aerogenes NCTC 418 and plasmid pRK2501 in Pseudomonas putida CYM 318. J Basic Microbiol. 30:57-61

Tsong, T. Y. 1991. Electroporation of cell membrane. Biophys.J. 60:297-306

Valentin, H. E. and D. Dennis. 1996. Application of an optimized electroporation procedure for replacement of the polyhydroxyalkanoate synthase I gene in Nocardia corallina. Can J Microbiol. 42:715-719

Wang, T. W. and Y. H. Tseng. 1992. Electrotransformation of Xanthomonas campestris by RF DNA of filamentous phage phi Lf. Lett Appl Microbiol. 14:65-68

Wards, B. J. and D. M. Collins. 1996. Electroporation at elevated temperatures substantially improves transformation efficiency of slow-growing mycobacteria. Fems Microbiol Lett. 145:101-105

Williams, S. T., M. E. Sharpe and J. G. Holt. 1989. Bergey's manual of systematic bacteriology in Williams & Wilkins, Baltimore.

Xie, T. D., L. Sun and T. Y. Tsong. 1990. Study of mechanisms of electric field-induced DNA transfection. I. DNA entry by surface binding and diffusion through membrane pores. Biophys. J. 58:13-19

Xie, T. D., L. Sun, H. G. Zhao, J. A. Fuchs and T. Y. Tsong. 1992. Study of mechanisms of electric field-induced DNA transfection. IV. Effects of DNA topology on cell uptake and transfection efficiency. Biophys J. 63:1026-1031

Xie, T. D. and T. Y. Tsong. 1990. Study of mechanisms of electric field-induced DNA transfection. II. Transfection by low-amplitude, low-frequency alternating electric fields. Biophys. J. 58:897-903

Xie, T. D. and T. Y. Tsong. 1992. Study of mechanisms of electric field-induced DNA transfection. III. Electric parameters and other conditions for effective transfection. Biophys J. 63:28-34

Yamamoto, N. and T. Takano. 1996. Isolation and characterization of a plasmid from Lactobacillus helveticus CP53. Biosci Biotechnol Biochem. 60:2069-2070

Yeung, M. K. and C. S. Kozelsky. 1994. Transformation of Actinomyces spp. by a gram-negative broad-host-range plasmid. J Bacteriol. 176:4173-4176

Zealey, G., M. Dion, S. Loosmore, R. Yacoob and M. Klein. 1988. High frequency transformation of Bordetella by electroporation. Fems Microbiol Lett. 56:123-126

Zimmermann, U. 1982. Electric field mediated fusion and related electrical phenomena. Biochim. Biophys. Acta. 694:227-277

Zimmermann, U. and G. A. Neil. 1996. Electromanipulation of cells in CRC press, N Y, Boca Raton, Florida.

A Critical Introduction to the Technology of Bacterial Electrotransformation

JUSTIN TEISSIÉ

Introduction

The basic principle of electrotransformation is to apply electric pulses to a mixture of bacteria and plasmids in an aqueous solution. In most protocols, only one pulse is needed. This is obtained by applying a selected voltage U (volts) on two electrodes, the interelectrode space being filled with the mixture. In all experiments dealing with gene transfer in bacteria, the two electrodes are flat and parallel with a width d (cm). As a consequence, a field E (V/cm) is present between the two electrodes with a strength:

$$E = U/d \qquad\qquad (1)$$

To be effective for gene transfer in bacteria the field intensity must be high (several kV/cm). This is obtained by using either a high voltage or a narrow width. For reasons of cost, it is difficult to obtain very high voltages under easy to handle conditions. Narrow chambers must therefore be used. But for pipetting reasons, it is difficult to work with electrode width, d, smaller than 0.1 cm.

Another parameter must be taken into account, the current I (Amp), which flows across the suspension when the voltage is present. There is a linear relationship between voltage and current, which is given by Ohm's law:

$$U = R_S\, I \qquad\qquad (2)$$

Justin Teissié, IPBS CNRS (UPR 9062), 118 Route de Narbonne, Toulouse, 31062, France (*phone* +33-561-33-58-80; *fax* +33-561-33-58-60; *e-mail* justin@ipbs.fr) where Rs (ohm, Ω) is the resistance of the sample. When the electrodes are flat and parallel, as is the case for the pulsing chambers which are commonly used, Rs is given by:

$$Rs = d^2/(\Lambda\ Vol) \tag{3}$$

where Λ is the conductivity of the sample (cm/Ω) and Vol is the volume of the sample (cm^3).

Rs must be large to limit the current which is delivered to the sample. As just reported, d is small. Therefore, from Eq(3), Λ and Vol must be small. In most experiments the sample volume is 40 µL or less. Low values of Λ are obtained by using glycerol solutions in water or buffered solutions where the osmolarity is balanced by the addition of sucrose. Problems are clearly present when using the second kind of pulsing buffers, because ions are present. Ions increase the conductivity. They must be eliminated or present under controlled conditions. A key point is therefore that the bacterial suspension must be washed several times in order to eliminate all the contamating salts coming from the culture medium. A similar problem is associated with the DNA solution. This is a problem when working with ligation products. As will be described, if the conductivity of the mixture is high, the resulting sample resistance is low and problems occur.

Two technologies are used to obtain a high voltage: capacitor discharge or square wave generator.

Definition of the pulse duration

In the capacitor discharge system, a high voltage power supply, which supplies a low current, is used to load a capacitor C (farad) to a preset voltage. The power supply is disconnected; the capacitor is then connected to the pulsing chamber, which is then the discharge resistor of the capacitor. All these connections and disconnections are of course within the built-in circuit and triggered by just pressing a switch. The voltage, which is present between the two electrodes, decays with a rate constant K

$$K = 1/(RC) \tag{4}$$

For technical reasons (definition of the sample volume, wetting of the electrodes), it is indeed very difficult to get controlled values for Rs in such a simple set-up. Most commercial electropulsators are built with a resistor R_L in parallel with the sample. As Rs is supposed to be large (see above), R_L is chosen to be much smaller (it is set at 200 Ω in most machines). When this is true, the decay time of the voltage, τ (s), i.e. of the field applied on the sample, is given by the reciprocal of the rate constant, which is now 1 / (R_L C). The decay time is then

$$\tau = R_L C \tag{5}$$

τ is the time when the field intensity has decayed by about 67%.

To limit the current flowing across the sample, another resistor Rsc is present in series between the capacitor and the sample.

- Low cost electropulsators are built with a given pair R_L and C in such a way as to obtain a value of τ about 5 ms.

- More sophisticated systems have a bank of capacitors and resistors in order to give more flexibilities in the choice of τ.

- τ is a critical parameter in Electropulsation. Flexibility in its choice is of prime importance when a new protocol is under design.

A major drawback is present in the definition of τ, given in Eq(5). The assumption is that:

$$Rs > R_L \tag{6}$$

As described above, this condition is not fulfilled if Rs is small. This is the case when:

- the bacterial suspension is not carefully washed

- the plasmid preparation is not free of ions

- the choice of the pulsing buffer is wrong (high salt content).

To avoid these problems, high priced electropulsators are able to measure the sample resistance Rs before the pulse is applied to warn the user if its value is amazingly low.

Nevertheless, even if the sample preparation is carefully controlled, a problem may be present when using highly concentrated bacterial suspensions. This appears as a technical advantage because increasing the bacterial concentration increases the total number of transformants for a given volume and a given amount of added DNA. A limit is nevertheless present. It was shown that as soon as the field triggers the cell membrane permeabilization, a leakage of the cytoplasmic content is present. When the concentration of bacteria is very high, even when the preparation has been carefully washed out and a buffer with a very low ionic content is used, the leak of the cytoplasmic ions induces a sharp drop in the sample resitance Rs. A decrease by 3 orders of magnitude was reported in the case of experiments with E coli. Eq(6) is not valid anymore during the pulse and as a consequence the definiion of τ given in Eq (5) is no longer valid. As Rs decreases during the pulse, the decay of the field is not exponential anymore. A practical consequence is that the field is applied on the sample during a duration much shorter than predicted from Eq(5). No control of the pulse duration is present. As in most systems, the decay of the field is not recorded, so it is impossible to know what was done. Erratic results are obtained in many cases.

Square waved pulses are obtained with a more elaborate technology. A high voltage high power generator designed to provide a high current is driven by an electronic clock to deliver the voltage on the electrodes during a calibrated duration T (s). The field is supposed to remain constant during the pulse. These electropulsators treat the bacteria under controlled definitions of both the field strength and pulse duration. The two parameters are independently adjustable. This gives more flexibility when establishing new protocols.

For technical reasons the design of such electropulsators is more difficult than for capacitor discharge systems. As a consequence very few are available on the market providing conditions suitable for gene transfer in bacteria.

The major problem is again linked to the sample resistance Rs, which in their case is directly connected to the generator. If Rs is low, the current must be high and cannot be delivered for a long duration, because the electical power P (watt):

$$P = UI \qquad (7)$$

must be very high. This cannot be obtained at a moderate cost.. With low values of Rs, the voltage cannot keep a steady value and a pseudo exponential decay is observed.

As described for the capacitor discharge technology, the following limits are present:

- the bacterial suspension must be carefully washed

- the DNA preparation must be ion free

- the content of the pulsing buffer must be poor in ions

- the bacterial concentration must be kept rather low ($5 \cdot 10^8$ / mL as a maximum for E. coli).

If these conditions are fulfilled, a good control of the pulse duration is obtained during the treatment.

Heating problems

As a current flows across the sample during the pulse, a Joule Heating effect is present. The electrical energy is converted into heat, which gives a temperature increase of the sample.

The electrical energy is stored in the capacitor

$$W = CU^2/2 \qquad (8)$$

Capacitor discharge system

It is dissipated in the sample (R_s) and in the parallel resistor R_L during the discharge. The contribution of R_{sc} is negligible. The part of this energy which is dissipated in the sample is:

$$Ws = RCU^2/(2\ Rs) \qquad (9)$$

where R is the equivalent resistance of R_s and R_L in parallel

$$R = Rs\ R_L/(Rs + R_L) \qquad (10)$$

If the sample is carefully prepared as described above, then

$$Rs > R_L \qquad (11)$$

and

$$Ws = R_L\ CU^2/(2\ Rs) \qquad (12)$$

Only a minute fraction of the energy is dissipated in the sample. This energy induces a temperature increase $\Delta\theta$ such as

$$WS = J\ \rho\ Vol\ \Delta\theta \qquad (13)$$

J being 4.18 J/cm^3

ρ being the calorific coefficient of the sample (assumed to be the one of water) The temperature increase is then

$$\Delta\theta = R_L \ CU^2/(2 \ Rs \ J \ \rho \ Vol) \tag{14}$$

The temperature increase is therefore strongly dependent on the R_L over Rs ratio. If a 40 µL sample is pulsed under 1000 V with a 25 µF capacitor, a condition which is routinely used, $\Delta\theta$ will increase linearly with the R_L / Rs ratio from less than 1°C when the ratio is 0.01 up to more than 70°C when it is 1. This means that if the sample is kept at room temperature before pulse application, its final temperature will be over 90°C.

This shows one of the advantages of keeping the sample on ice. But the pulsing chamber must be cooled down too, a feature which is not always available.

Again a careful control of the sample resistance is needed to obtain reliable electrotransformation conditions.

Nevertheless, problems may be present when highly concentrated bacterial suspensions are used. It was described above that due to the permeabilization associated leakage, the ionic content of the sample increases strongly during the pulse. As a consequence, Rs decreases. Even when present before the pulse, Eq(11) will not be obeyed when the leakage is present. A strong heating will result. The technical problems resulting from heating can be summarized as follows:

- sample denaturation

- sample boiling

This second consequence may give arcing of the sample and its associated damages to the electropulsator. Some protections are present on most commercial set-ups to limit the current (Rsc).

Square wave generator   The energy which is dissipated in the sample is then:

$$Ws = U^2T/Rs \tag{15}$$

The temperature increase is therefore

$$\Delta\theta = U^2T/(Rs \ J \ \rho \ Vol) \tag{16}$$

To reduce its extent under given pulsing conditions, it is clear that Rs must be as large as possible. Again the sample must be carefully prepared to avoid ionic contamination. One should notice that $\Delta\theta$ does not depend on the sample volume, Vol, for defined pulsing conditions (preset values of E and T). By taking into account the definition of Rs, Eq (16) can be rewritten as

$$\Delta\theta = E^2 \; T \; /(J \; \rho) \tag{17}$$

$\Delta\theta$ is a function of only the pulse duration and of the square of the applied field strength.

Nevertheless increasing the sample volume requires an increase in the energy which is delivered by the generator. This brings a limit in the experimental possibilities.

Electrodes

Most companies market cuvettes made of polycarbonate with aluminium electrodes, which are built in. The width between the electrodes can be chosen between 3 different values: 0.1, 0.2 and 0.4 cm, the narrower ones being suitable for use with bacteria because high field strengths can be obtained. The manufacturing process guarantees an exact tolerance in the gap width. They are sold sterile, individually wrapped and their electrode surface is carefully cleaned.

Single use cuvettes

Problems are nevertheless present:

- They are sold for single use. But as they are rather expensive, it is tempting to re-use them. As a strong current is present during the pulse, some surface damage is induced which may affect their further use.

- Some oxydation occurs on the electrode surfaces during storage. Aluminium oxyde affects the value of the voltage which is applied on the sample. A lower value than the one present on the electrodes is observed.

- Aluminium ions are released during the pulse. They affect cell viability and the stability of some biological molecules.

- It was shown that the energy which is dissipated is less than predicted, showing that erratic pulsing conditions are present.

The conclusion is then that these electrodes are convenient for routine experiments, where the protocol is well established and the number of transformants can be high. They do not appear very suitable when designing a new protocol with recalcitrant strains.

Stainless steel electrodes   Two companies propose electrodes which may be used several times. They can be coated with gold or platinium to improve their electrochemical properties. One problem in their use is that they must be washed and sterilized before use. This is obtained simply by a wash with 70% ethanol and rinsing with sterile water. This is the major problem in their use.

Minor troubles are:

- Release of ferric ions, but to a low extent under the conditions used in bacterial electrotransformation. It was suggested to cover them with a wet filter to prevent their diffusion in the sample volume.

- rusting of electrodes

It was checked that the voltage present on the sample was what was set on the electrodes, i.e. on the generator and / or capacitor.

Electropulsators

In the early days of bacterial electrotransformation, some groups designed their own set-ups. This is not a good approach nowadays. Several companies market specifically designed electropulsators. Their technical specifications are briefly reported in Table 1. Three categories can be discriminated.

Table 1.

A- Low cost Electropulsator

Company	Model	Power supply	Voltage range (V)	Decay time (ms)	cuvette
BTX	Transporator	not provided	0-2500	5	single use
BTX	Transporator plus	provided	0-2500	5	single use
BTX	ECM 395	provided	300-3000	5	single use with PEP

Table 1. Continoued

A- Low cost Electropulsator

Company	Model	Power supply	Voltage range (V)	Decay time (ms)	cuvette
Eppendorf	Electroporator 2510	provided	200-2500 preset 1800 or 2500	5	single use
EC Apparatus corp	EC100	provided	preset 1800 or 2500	5	single use
Biorad	E coli pulser	provided	200-2500 preset 1800 or 2500	5	single use
Equibio	Easy ject basic	provided	preset 2500	5	single use
Equibio	Easyject Prima	provided	preset 1800 or 2500	5	single use

B Capacitor discharge system

Company	Model	Power supply	Voltage range (V)	Resistors	Capacitors
Invitrogen	Electroporator II	not included	150-1500	70 ohms to infinite	50 and 71 mF
Biorad	Gene pulser II pulse controller II (1)	provided	50-2500	6 different values	25 mF
Biorad	Gene pulser II pulse controller plus (1)	provided	50-2500	11 values	25 mF
BTX	ECM 600 (2)	provided	50-2500	10 values	126 values
Equibio	Easyject one	provided	100-2500	10 values	0.5 and 2.5 mF
Equibio	Easyject plus	provided	100-3500	10 values	0.5 and 2.5 mF
Equibio	Easyject optima (3)	provided	20-2500	not described	5 to 1500 mF

(1) Technical bulletins are available

(2) A data base support is provided

(3) Software control

C- Square Wave Electropulsator

Table 1. Continoued

Company	model	Voltage range (V)	Duration range (ms)	cuvette
BTX	T820 (1)	100-3000	0.005-0.099	any kind
Jouan	PS15 (2)	100-1500	0.005-24	any kind

(1) A data base support is provided

(2) A high frequency of pulse repetition is available

Low cost systems They use the capacitor discharge methodology. They are pre set at one given voltage and one decay time (5 ms). But this second feature is true only when the sample is carefully prepared to avoid ionic contamination and a moderate bacterial concentration is treated.

They use single use aluminium cuvettes, with in many cases a built-in cuvette holder. Most of them have their own high voltage power supply.

They appear suitable only for routine experiments. Their name indeed refers often to *E. coli*.

Capacitor discharge systems The voltage can be adjusted by proper settings on the power supply. The decay time is controlled by a bank of capacitors and discharge resistors, giving several values for R_L.

Some of these machines allow a predetermination of the sample resistance before the pulse, to be sure that proper conditions are present.

Safety devices supposed to prevent arcing are present (Rsc). They use single use aluminium cuvettes.

Square wave pulsators Only two models are marketed and one of them delivers short pulses which are not very suitable for experiments to obtain gene transfer and expression in bacteria.

The pulsing chamber is always separated from the generator.

They offer the highest flexibility in the experimental procedure.

References

Eynard N., Sixou S., Duran N. and Teissie J. (1992) Fast kinetics studies of E.coli electrotransformation Eur. J. Biochem; 209, 431-436

Loomis-Husselbee J.W., Cullen P.J., Irvine R.F. and Dawson A.P. (1991) Electroporation can cause artefacts due to solubilization of cations from the electrode plates Biochem. J. 277, 883-885

Pliquett U., Gifte A. and Weaver J.C. (1996) Determination of the electric field and anomalous heating caused by exponential pulses with aluminium electrodes in electroporation experiments, Bioelectrochem. Bioenerg. 39, 39-53

Suppliers

Biorad
Hercules, Ca 94547, USA
Phone 1 (510) 741 1000
Fax 1 (510) 741 5800
http://www.bio-rad.com

BTX
San Diego, Ca 92121-1334, USA
Phone 1 (619) 597 6006
Fax 1 (619) 597 9594
http://www.genetronics.com

Eppendorf
Madison, WI 53711-1082, USA
Phone 1 (608) 231 1188
Fax 1 (608) 231 1339

Equibio
Boughton Monchelsea, Kent, UK
Phone 44 (0) 1622 746300
Fax 44 (0) 1622 747060
Email 101552.300@compuserve.com
http://ourworld.compuserve.com/homepages/equibio/

Invitrogen
NV Leek, 9351, The Netherlands
Phone 31 (0) 594 515 175
Fax 31 (0) 594 515 312
Email tech_service@invitrogen.nl
http://www.invitrogen.com

Jouan
St Herblain, 44805, France
Phone 33 (0) 2 40 16 8000
Fax 33 (0) 2 40 94 7016
Email jouan@jouan.worldnet.fr

Chapter 3

Electrotransformation of *Escherichia coli*

NATALIE EYNARD and JUSTIN TEISSIÉ

Introduction

E. coli is the most popular bacteria in laboratories of bacterial genetics and biotechnology. Getting DNA transfer and expression in such a system is required in many experiments. Electropulsation is now known as being a highly efficient approach. Its use is very easy (just mix plasmid and bacteria, then pulse) and not time consuming. Very simple electropulsators can be used (*E. coli* pulser). There is no need to induce a wall modification in the case of *E. coli*. No major improvements are found after EDTA and lysozyme treatment. Most manufacturers provide protocols for *E. coli* electrotransformations, which can be downloaded from the WEB.

Materials

Strains

Transformation of *E. coli* cells is now routinely used with an efficiency of 10^9 transformants per µg of DNA. The increase of this efficiency to 10^{10} transformants per µg of DNA is important when transformation is used for constructing a cDNA library of rarely expressed genes. But nevertheless the cell survival must remain high after pulsing. *E coli* can be easily electro-transformed, but the efficiency remains strongly strain dependent. The highly electrotransformable strains are DH5α, DH1,

✉ Natalie Eynard, IPBS CNRS (UPR 9062), 118 Route de Narbonne, Toulouse, 31062, France (*phone* +33-561-33-58-80; *fax* +33-561-33-58-60; *e-mail* eynard@ipbs.fr)
Justin Teissié, IPBS CNRS (UPR 9062), 118 Route de Narbonne, Toulouse, 31062, France

JM109, XL-1 and Blue MR (Sheng, et al., 1995) compared to the less transformable strains as HB 101, NM538 (Tung and Chow, 1995), BL21 (DE3) (data not shown), or AB1157 (Sixou, et al., 1991). One should notice that strains which were optimized for chemical transformation are easily electrotransformed. Finally some manufacturers provide electro-competent cell kit (cultured, washed and frozen bacteria) as Bio-rad (Hercules, Ca) or Invitrogen (Gronigen, NL) or Genetronix (San Diego, CA).

Gentronix: www.genetronics.com/btx

Eppendorf: www.eppendorf.com/eppendrf

Biorad: www.bio-rad.com

Invitrogen: www.invitrogen.com

Culture media   – LB medium
- To 950 ml of deionized H_2O add:
- 5 g yeast extract (Difco, Detroit, MI)
- 10 g bactotryptone (Difco)
- 5 g/l NaCl (Sigma, St Louis, MO)
- Adjust to 1 liter
- Adjust to pH 7.4 with 1N NaOH
- Autoclave to sterilize

– YENB
- To 950 ml of deionized H_2O add:
- 7.5 g yeast extract (Difco)
- 8 g Nutrient broth (Difco)
- Adjust to 1 liter
- Adjust to pH 7.4 with 1N NaOH
- Autoclave to sterilize

– SOB without salt
- To 950 ml of deionized H_2O add:
- 5 g yeast extract (Difco, Detroit, MI)
- 20 g/l bactotryptone (Difco)
- Adjust to 1 liter
- Adjust to pH 7.4 with 1N NaOH
- Autoclave to sterilize

– SOC
- To 950 ml of deionized H_2O add:
- 5 g yeast extract (Difco, Detroit, MI)
- 20 g/l bactotryptone (Difco)
- 0.5g NaCl (Sigma)
- 0.2g KCl

- Adjust to 1 liter
- Adjust to pH 7.4 with 1N NaOH.
- Autoclave to sterilize
- After cooling at room temperature add:
- 20 ml of a sterile solution of 1 M of glucose
  (sterilized by autoclaving)
- Just before use add:
- 5 ml of a sterile solution of 2 M of $MgCl_2$
  (sterilized by autoclaving),
- 5 ml of a sterile solution of 2 M of $MgSO_4$
  (sterilized by autoclaving).

All liquid media can be stored at room temperature.

Solid media are obtained by addition of agar (15 g/l) before autoclaving. Store at 4°C.

For selective media, stock solution of antibiotics is added after autoclaving and cooling at room temperature.

Pulsing buffer: **Buffers**

- PB 1:
  - To 950 ml of deionized H_2O add:
  - 1 ml of a sterile solution of 1 M of Tris, pH 7.4
  - 270 to 500 ml of 1 M Sucrose (Sigma)
  - Autoclave to sterilize
  - after cooling at room temperature add:
  - 1 ml of a sterile solution of 1 M of $MgCl_2$
    (sterilized by autoclaving)
- PB 2:
  - Sterile water with 10% (w/v) glycerol (Sigma)
    (sterilized by autoclaving)
  - In some cases sucrose (270 mM) (sterilized by autoclaving) or DMSO (7% v/v, (Chuang, et al., 1995)) is preferred, sterile filtered on filter units (0.2 μm, Nalgene).
- TE buffer:
  - To 90 ml of deionized H_2O add:
  - 1 ml of a sterile solution of 1 M of Tris, pH 7.4
  - 20 μl of 0.5M EDTA (Sigma).
  - Adjust to 100 ml.

Procedure

Preparation of "electro-competent" cells

1. A fresh 37°C-overnight culture is diluted from 1:1000 to 1:50 into 1 liter of fresh medium (LB medium in general).
   In the case of a generator producing exponentially decaying pulses, the efficiency of the applied pulse depends on the sample resistance. To keep the resistance at high value, it is important to remove all salts from the cell suspension. In these cases (Bio-Rad, BTX, Eppendorff, ...) a rich growth medium, which is not supplemented by salt can be used such as YENB (Sharma and Schimke, 1996) or SOB (Sheng, et al., 1995).

2. Cells are grown classically with shaking (200-300 RPM) at 37°C.
   It has been shown that growth at low temperature increases transformation frequency for some strains (18°C, (Chuang, et al., 1995)).

3. Culture is harvested in the early exponential phase (A_{650nm} of 0.4) by centrifugation (4°C, 6000g, 10 min).
   E. coli in stationnary phase are not easily transformed, a loss in viability is observed.

4. After removing the medium, the pellet is washed twice in 1 liter of cold (4°C) pulsing buffer PB and centrifuged again (twice is sufficient in general but some authors make a third wash).

5. Cells are finally concentrated 100 fold in 10 ml of cold PB (approximatively 10^{10} CFU/ml).
   For the apparatus which delivers square wave pulses, the PB could contain some salts. Presence of magnesium seems to enhance the cell viability of fragile strains and sucrose creates an increase of the external osmolarity which protects the cell during the membrane permeabilization (Eynard, et al., 1992). 0.5 M sucrose gives an optimum in osmolarity (PB 1). As for the choice of culture medium, in the case of a generator producing exponentially decaying pulses, the resistance in the pulsing medium has to be as high as possible, and the classical PB contains only glycerol or DMSO (PB 2).

6. Cells prepared in this way can be used directly or be frozen (by adding 10% glycerol in the case of the PB 1, or directly with PB 2).

7. Freezing procedure: dispatch 500 µl aliquots in cryo-tubes (Nunc, Kamstrup, DK) and store at - 70°C.
   Cells stored in this way can be used at least one year after freezing. It is important that all steps from the first centrifugation to the the concentration in PB were made at 4°C.

8. Before use, thaw frozen samples slowly on ice (washed in same volume of the PB when using the PB1).

Preparation of cells-DNA mixture

DNA must be prepared with minimum level of salt. Plasmids can be purified either by classical methods such as triton-lysozyme or alkaline lysis (Davis, et al., 1986) or by using a Quiagen column (Hilden, DE). Purified plasmid can be resuspended in low ionic strength buffer (TE) or simply in water.

1. Mix an aliquot of 30 to 40 µl of cells with 1 to 5 µl of DNA. Mixture of cells and plasmid can be done directly in the electropulsation cuvette or in a plastic tube. After a good mix no incubation on ice is needed before applying the pulse. The total volume of the cell DNA mixture depends on the cuvette used for the experiment. Approximatively 10^9 cells are electropulsed in an experiment and the transformation efficiency is proportional to the final DNA concentration over a wide range (at least 0.1 to 5 µg/ml). The plasmid used is preferentially supercoiled or circular (linearised forms give few transformants probably because of their degradation in the cytoplasm) (Xie, et al., 1992, Xie and Tsong, 1992).

2. Put the mixture in the cuvette
   When filling the cuvette, check that no bubbles are present in the suspension. They may induce arcing when high fields are triggered. This is a problem when using the 0.1 cuvettes.

Electropulsation

The electrical parameters depend on the generator used for the experiments.

- With a square wave pulses delivering generator apply 1 pulse of several milliseconds (between 3 and 7 ms) duration at 6-9 kV/cm (the maximum value is obtained with 7 ms between 6 and 9 kV/cm (Sixou, et al., 1991)).
- With an exponential generator, a pulse controller (i.e. resistor in parallel to the cuvette in the discharge circuit) must always be used. Apply an exponential decay pulse of 2.5 kV, R= 180-200 Ω, C = 25 -40 μF. The time constant obtained in this case will be about 4 - 7 ms and the initial field strength of 12.5 or 25 kV/cm for respectively a 0.2 or 0.1 cm cuvette.

The initial value of the applied field is always higher with a capacitor discharge system than the mean value set for a square wave generator. For exponential decaying pulse, the field applied is not over the critical value needed for transformation during all the pulse. Salts can be present when using a square wave pulse because they do not affect the pulse duration but a high salt concentration should be avoided because it brings a strong temperature increase.

Expression of transformants

1. Immediately after pulsing, the electropulsed volume is mixed with 1 ml of culture medium at room temperature.
   Mix by shaking the cuvette. This is very important when using the 0.1 cm cuvette to obtain an homogeneous suspension.
   In general LB can be used but higher efficiency is obtained with SOC medium. – Then the mixture is put on a sterile tube and incubated for 1 h at 37°C (shaking is not necessary).

2. Cells are then plated on solid medium containing the appropriate selecting agent and incubated at 37°C.
   The number of resulting transformants depends on the quantity of DNA put in the cuvette, the dilution used and the volume of aliquot spread on the plate (classically 100 µl, but 300 or 400 µl can be used when expected efficiency is low).

The efficiency E = transformants on the plate x dilution x (10^{-3} / volume (liter) spread on the plate) x (10^{-6} / quantity of DNA (gram) in the cuvette).

10^{-3} is for the total volume of electropulsed cells (approximatively 1 ml with the culture medium) 10^{-6} represents the 1 µg DNA used for expression of efficiency.

3. Colony forming units (CFU) can be counted after an overnight or 24 h incubation.

4. To estimate the cell survival, spread an 100 µl aliquot on non selective medium with a preliminary dilution 10^{-5} or 10^{-6} (if 10^9 cells were pulsed).

    The cell viability is strongly affected by the electrotransformation procedure. Using drastic conditions gives a high loss of viability but a high number of transformed cells in the surviving bacteria.

▓ References

Chuang, S. E., A. L. Chen and C. C. Chao. 1995. Growth of E. coli at low temperature dramatically increases the transformation frequency by electroporation. Nucleic Acids Res. 23:1641

Davis, F. G., M. D. Dibner and J. F. Battey. 1986. Basic methods in molecular biology in Elsevier Science, Amsterdam

Eynard, N., S. Sixou, N. Duran and J. Teissie. 1992. Fast kinetics studies of Escherichia coli electrotransformation. Eur J Biochem. 209:431-436

Sharma, R. C. and R. T. Schimke. 1996. Preparation of electrocompetent E. coli using salt-free growth medium. Biotechniques. 20:42-44

Sheng, Y., V. Mancino and B. Birren. 1995. Transformation of Escherichia coli with large DNA molecules by electroporation. Nucleic Acids Res. 23:1990-1996

Sixou, S., N. Eynard, J. M. Escoubas, E. Werner and J. Teissie. 1991. Optimized conditions for electrotransformation of bacteria are related to the extent of electropermeabilization. Biochim Biophys Acta. 1088:135-138

Tung, W. L. and K. C. Chow. 1995. A modified medium for efficient electrotransformation of E. coli. Trends Genet. 11:128-129

Xie, T. D., L. Sun, H. G. Zhao, J. A. Fuchs and T. Y. Tsong. 1992. Study of mechanisms of electric field-induced DNA transfection. IV. Effects of DNA topology on cell uptake and transfection efficiency. Biophys J. 63:1026-1031

Xie, T. D. and T. Y. Tsong. 1992. Study of mechanisms of electric field-induced DNA transfection. III. Electric parameters and other conditions for effective transfection. Biophys J. 63:28-34

Transformation of *Bacillus subtilis* PB1424 by Electroporation

PATRIZIA BRIGIDI, MADDALENA ROSSI and DIEGO MATTEUZZI

Introduction

Bacillus subtilis is a potential model system for biochemical and genetic engineering of Gram-positive bacteria and represents an attractive host for the extracellular production of foreign proteins. Several species of the *Bacillus* genus have been successfully electroporated: *B. subtilis* (Brigidi et al 1990, Ohse et al 1995, Ohse et al 1997), *B. thuringiensis* (Bone and Ellar 1989, Shurter et al 1989), *B. cereus* (Belliveau and Trevors 1989, Shurter et al 1989), *B. amyloliquefaciens* (Vehmaapera 1989), *B. stearothermophilus* (Narumi et al 1992), *B. sphaericus* (Taylor and Burke 1990), *B. brevis* (Takagi et al 1989), *B. licheniformis spea* (Brigidi et al 1991) and *B. anthracis* (Quinn and Dancer 1990). Although the plasmid transformation of *B. subtilis* can be accomplished either with competent cells or by protoplasts in presence of polyethylene glycol, electroporation is sometimes the preferred method because it is less tedious and time consuming.

✉ Patrizia Brigidi, University of Bologna, Department of Pharmaceutical Sciences, Interdepartmental Center for Biotechnology, Via Belmeloro 6, Bologna, 40126, Italy (*phone* +39-51-2099727; *fax* +39-51-2099734; *e-mail* patbri@alma.unibo.it)

Maddalena Rossi, University of Bologna, Department of Pharmaceutical Sciences, Interdepartmental Center for Biotechnology, Via Belmeloro 6, Bologna, 40126, Italy

Diego Matteuzzi, University of Bologna, Department of Pharmaceutical Sciences, Interdepartmental Center for Biotechnology, Via Belmeloro 6, Bologna, 40126, Italy

Materials

Bio-Rad Gene Pulser™ coupled to a Bio-Rad Pulse Controller   **Equipment**
and 0.2 cm gap disposable cuvettes (Bio-Rad).

Procedure

Preparation of electrocompetent cells

1. Streak a LB plate with cells of strain PB1424 and grow for 12-16 h at 37°C.

2. Inoculate 20 ml of LB medium (300-ml baffled culture flask) with a large clump of cells and grow the culture in a gyratoty shaker (200 rpm) for 2 h at 37°C.

3. Inoculate 250 ml of prewarmed LB medium (1 l baffled culture flask) with 2.5 ml of the 2 h growth culture. Grow the culture with shaking (200 rpm) at 37°C to an O.D.$_{600}$ of 0.6-0.7, corresponding to a mid-log phase.

4. Chill the cells on ice for 15 min.

5. Transfer the culture to a sterile ice-cold 250 ml centrifugation bottle and collect the cells by centrifugation at 5000 x g for 15 min at 5°C.

6. Decant the supernatant fluids and suspend the cells in 250 ml of sterile, ice-cold distilled water. Vigorously shake the bottle until the cells are suspended. Collect the cells as in step # 5.

7. Suspend the cells in 250 ml of sterile, ice-cold distilled water as in step # 6 and collect the cells as in step # 5.

8. Decant the supernatant fluids, suspend the cells in 20 ml of sterile, ice-cold 10% glycerol in water, transfer to a 50 ml centrifuge tube, and collect the cells as in step # 5.

9. Repeat step # 8.

Note: Keep the cells on ice during steps # 5-9.

10. Suspend the cell pellet in 1 ml of sterile, ice-cold 10% glycerol. Dispense 40 µl aliquots of cell suspension into cold sterile 1.5 ml centrifuge tubes and store at -80°C. Before freezing,

dilute the fresh cell suspension in LB medium, and plate 0.1 ml samples of the 10^{-7} dilution, in quadruplicate, on LB plates and incubate the plates overnight at 37°C. The cell density should be at least 2×10^{10} CFU/ml. Electrocompetent cells stored at -80°C exhibit the same transformation efficiency for at least 4 months.

Electrotransformation

1. Hand thaw the cells and immediately place the tubes on ice. Place the electrotransformation cuvettes (0.2 cm gap) on ice.

2. To the vial containing 40 µl of cell suspension add 1 to 2 µl of plasmid DNA in TE buffer pH 8 (0.25-0.5 µg DNA) and hold on ice for 1 min. Set the Bio-Rad Gene Pulser™ and the Pulse Controller at 25 µF capacitance, 1.4 kV voltage and 200 Ohms resistance: these conditions generate time constants ranging from 4 to 5 msec.

3. Quantitatively transfer the cell-DNA mixture to a cold electrotransformation cuvette. Examine the cuvette to ensure that the liquid completely fills the bottom and that there are no bubbles (can result in ARC-ing) and immediately place the cuvette into the cuvette chamber of the Bio-Rad Gene Pulser™.

4. Immediately after the pulse add 1ml of LB and transfer the electrotransformed cell suspension to a sterile polypropylene tube.

5. Incubate the tube in a gyratory shaker (200 rpm) for 1 h at 37°C.

6. Plate samples (0.1 ml) of the undiluted electrotransformed culture on the appropriate selective medium and incubate the plates at 37°C for 16-18 h.

Results

Electroporation can be used to transform *B. subtilis* strain PB1424 at efficiencies of 10^3-10^4 transformants/μg of DNA depending on the replicon used. Although the efficiency of transformation of the competent cells of this strain is higher, the electroporation is a much more simple and rapid method. The effect of applied voltage on transformation efficiencies is the most significant parameter. Voltage optima round 1.2-1.4 kV, from 1.4 kV up to 2 kV efficiency gradually decreases and at electric pulses higher than 2 kV no transformants are selected. Cell viability of *B. subtilis* PB1424 is not significantly affected by increased electric field strength in the voltage range tested, the degree of cell death is around 80-95%. A marked increase in transformation efficiency is noted with mid-log phase cells at high cell concentrations (10^{10} CFU/ml), as the culture begins to approach its stationary phase, transformation efficiency decreases. Furthermore, the transformation efficiency decreases with increases of size of the DNA plasmid, however 10^3 transformants/μg DNA are obtained routinely by using plasmids with a large molecular size of about 10-12 kb.

The procedure described has proved to be effective for transferring plasmid DNA into intact cells of *B. licheniformis* 5A24 (Bacillus Genetic Stock Center, Ohio State University, Columbus, Ohio), with transformation efficiencies higher than those obtained for *B. subtilis* (10^4-10^6 transformants/μg of DNA). It is noteworthy that strain-to-strain variability is very common when electrotransforming *Bacillus* spp. Furthermore, use of ligated DNA for electroporation results in low transformation efficiencies.

References

Belliveau BH, Trevors JT (1989) Transformation of *Bacillus cereus* vegetative cells by electroporation. Appl Environ Microbiol 55:1649-1652

Bone EJ, Ellar DJ (1989) Transformation of *Bacillus thuringiensis* by electroporation. FEMS Microbiol Lett 58:171-178

Brigidi P, De Rossi E, Bertarini ML, Riccardi G, Matteuzzi D (1990) Genetic transformation of intact cells of *Bacillus subtilis* by electroporation. FEMS Microbiol Lett 67:135-138

Brigidi P, De Rossi E, Riccardi G, Matteuzzi D (1991) A highly efficient electroporation system for transformation of *Bacillus licheniformis*. Biotechnol Techniques 5:5-8

Narumi I, Sawakami K, Nakamoto S, Nakayama N, Yanagisawa T, Takahashi N, Kihara H (1992) A newly isolated *Bacillus stearothermophilus* K1041 and its transformation by electroporation. Biotechnol Techniques 6:83-86

Ohse M, Takahashi K, Kadowaki Y, Kusaoke H (1995) Effects of plasmids size and several other factors on transformation of *Bacillus subtilis* ISW1214 with plasmid DNA by electroporation. Biosci Biotechnol Biochem 59:1433-1437

Ohse M, Kawade K, Kusaoke H (1997) Effects of DNA topology on transformation efficiency of *Bacillus subtilis* ISW1214 by electroporation. Biosci Biotechnol Biochem 61:1019-1021

Quinn CP, Dancer BN (1990) Transformation of vegetative cells of *Bacillus anthracis* with plasmid DNA. J Gen Microbiol 136:1211-1215

Shurter W, Geiser M, Mathe D (1989) Efficient transformation of *Bacillus thuringiensis* and *Bacillus cereus* via electroporation: transformation of acrystalliferous strains with a cloned delta-endotoxin gene. Mol Gen Genet 218:177-181

Takagi H, Kagiyama S, Kadowaki K, Tsukagoshi N, Udaka S (1989) Genetic transformation of *Bacillus brevis* with plasmid DNA by electroporation. Agric Biol Chem 53: 3099-3100

Taylor LD, Burke WF (1990) Transformation of an entomopathic strain of of *Bacillus sphaericus* by high voltage electroporation. FEMS Microbiol Lett 66:125-128

Vehmaapera J (1989) Transformation of *Bacillus amyloliquefaciens* by electroporation. FEMS Microbiol Lett 61:165-170

Suppliers

Bio-Rad Laboratories
200 Alfred Nobel Drive
Hercules, CA 49547
USA
tel_ +510-741-1000 or +1-800-424-6723
fax_+510-741-5800 or +1-800-879-2289
telex 335-358

Part II

Biotechnology and Food Technology

Clostridium in Biotechnology and Food Technology

HANS P. BLASCHEK, JOSEPH FORMANEK and C.K. CHEN

Introduction

The genus *Clostridium* contains widely diverse species which have been shown to be associated with foodborne illness as well as solvent production (Blaschek, 1999). *C. perfringens* is representative of a clostridial species which is associated with food poisoning, while non-toxinogenic *C. beijerinckii* has important industrial and biotechnological potential because of its ability to produce acetone and butanol during fermentation of carbohydrates. The development of genetic systems for the clostridia was recently reviewed (Blaschek and White, 1995). Electroporation induced transformation has been shown to have considerable potential for introducing DNA into a wide variety of gram$^+$ and gram$^-$ bacterial species (Dower et al., 1992) and electroporation protocols are available for previously non-transformable clostridial strains (Chen et al., 1996). However, the transformation efficiencies appear to vary greatly among different species and strains. It has been suggested that DNase and restriction enzyme activity are common barriers to successful transformation of the clostridia. The cell wall may also represent a barrier to DNA uptake in *C. perfringens* (Kim and Blaschek, 1989). As we learn more about mechanisms and factors involved in trans-

✉ Hans P. Blaschek, University of Illinois, Department of Food Science and Human Nutrition, 1207 W. Gregory Drive MC-630, Urbana, Illinois, 61801, USA (*phone* +01-217-333-8224; *fax* +01-217-244-2517; *e-mail* blaschek@uiuc.edu)
Joseph Formanek, Griffith Laboratories, Central Research Department, One Griffith Center, Alsip, Illinois, 60803, USA
C.K. Chen, Southern Research Institute, Drug Discovery Program, 2000 Ninth Avenue, S., Birmingham, Alabama, 32555, USA

forming various bacteria, protocols for overcoming these obstacles will likely emerge. Electroporation protocols have been developed for the clostridia which are based on using polyethylene glycol (PEG) in the shocking menstruum. During electroporation-induced transformation, PEG may limit cytoplasmic leakage and, via exclusion of the aqueous phase, help to push the plasmid DNA into the cell and consequently increase cell viability and transformation efficiency.

Materials

Equipment Bio-Rad Gene Pulser^{TM} with Pulse controller or Bridge Muffy Electroporator$^{®}$

Media – TGY medium
– 3% trypticase
– 2% glucose
– 1% yeast extract
– 0.1% cysteine-HCl

Subprotocol 1
Electroporation-Induced Transformation
of *Clostridium perfringens*

Procedure

1. An early stationary phase *C. perfringens* cell culture (OD_{600} = 1.2) grown in TGY medium is harvested by centrifugation at 10,000 x g for 10 min. The cell pellet is washed once with 10% polyethylene glycol 8000 (PEG, Sigma) electroporation solution.

2. The washed cell pellet is suspended in 1/20 volume of 10% PEG solution to give ca. 10^8 to 10^9 cells per ml.

3. Cell suspension (0.8 ml) is mixed with 10-20 µl of plasmid DNA (ca. 1 µg) in a 0.4 cm electroporation cuvette without a pre-shock incubation.

4. Electroporation is performed using the Bio-Rad Gene Pulser™ with Pulse controller set at 25 μF, 1000 Ω, and 2.5 kV or the Bridge Muffy electroporator® set at 20 μF, 1000 Ω, and 2.5 kV.

5. The electroporated cell suspension is incubated on ice for 10 min and diluted into 9 volumes of TGY medium (1:10 dilution) and incubated for 3 hr at 37 °C anaerobically.

6. The cell culture is plated onto selective TGY plates containing 25 μg/ml of appropriate antibiotics and incubated overnight at 37 °C anaerobically. Cell recovery rate may be determined by plating cells on TGY minus antibiotics.

Results

The efficiency of electroporation-induced transformation is highly strain-dependent. As a result, it may be necessary to develop a specific set of electroporation parameters for a particular strain. Using the electroporation protocol described above, *C. perfringens* strain 13A was transformed with plasmid pGK201 at a transformation efficiency of nearly 10^9 transformants per μg plasmid DNA. This is at least 4-orders of magnitude higher than that observed for any other *C. perfringens* strain (Table 1). Pulse duration time and survival rate can be used as indicators for optimization of electroporation parameters. The pulse duration time can be adjusted by changing the field strength, resistance or capacitance. We found that 6.25 kV/cm field strength, 1000 Ω resistance, and 20 or 25 μF capacitance settings gave the best results for *C. perfringens* strains 13A, 3624A, 3626B and 3628C. Optimal transformation efficiency of *C. perfringens* strains was routinely obtained at 10 to 20 mS pulse duration time and 10 to 50% cell survival rate. Plasmids with methylated cytosine and adenine residues demonstrated better transformation efficiency of previously non-transformable *C. perfringens* type B strains than those without the modification (Chen et. al., 1996). The methylation protected the plasmids from degradation by endonuclease that is a common barrier for transformation.

Table 1. Electroporation-induced transformation of *C. perfringens* strains with plasmid pGK201

Bacterial strain	Duration time (mS)	Transformation efficiency[a]	Survival rate[b] (%)
C. perfringens 13A	16.2	8.1 x 108	28.5
C. perfringens 3624A	18.5	4.1 x 104	16.6
C. perfringens 3626B	14.1	2.5 x 102	41.6
C. perfringens 3628C	14.3	5.0 x 104	27.2

[a] Transformation efficiency is defined as transformants per μg DNA.

[b] Survival rate is defined as CFU recovered after electroporation divided by CFU recovered before electroporation.

▓ ▓ Troubleshooting

High field strength, greater than 6.25 kV/ cm, is commonly used for electrotransformation of gram + bacteria. However, this usually results in a low cell survival rate that, as a consequence, decreases or prevents transformation. Water quality is also extremely important in order to obtain reproducible transformation. We routinely use nanopure (Barnstead) water with a resistance of 17.3 m Ω-cm. Salts carried over from the plasmid DNA preparation may shorten the duration time of the shock and, thereby, negatively affect transformation efficiency. Moreover, when high field strength is used, a high salt concentration may cause arching in the electroporation cuvette which may reduce the cell survival rate.

The growth stage of the culture is important when preparing competent cells. With respect to *C. perfringens*, an early stationary phase culture has been shown to give the best results. Because of the presence of nucleases, the incubation of plasmid DNA with competent cells may result in degradation of plasmid DNA even when the incubation is carried out on ice. Therefore, it is advisable to eliminate a pre-shock incubation. It is to be noted that the erythromycin resistant gene, *ermAM*, which is widely used as a marker for *C. perfringens* vectors, is able to recombine within the chromosome of *C. perfringens* strains 13A and 3626B (Chen et. al., 1996). This recombinational ability may affect the ability to recover the vector in the autonomous state.

Subprotocol 2
Electroporation Induced Transformation
of *Clostridium beijerinckii*

▨▨ **Procedure**

1. Grow cells in TGY to an optical density at 600 nm of 1.2-1.4.

2. Centrifuge cells at 7000 x g for 10 minutes at 4 °C and discard the supernatant. Cells are washed twice by suspension in 1/2 volume electroporation buffer (10% PEG 8000; Sigma Chemical) and centrifuged as above.

3. Cells are suspended in 1/20 volume electroporation buffer and used immediately.

4. An appropriate amount of DNA (1 µg) plus cells is added to a sterile, pre-cooled electroporation (0.2 cm gapped) cuvette and shocked at 10 kV/cm, 25 µF capacitance and 600 Ω to infinite resistance for *C. beijerinckii* using the BioRad Gene Pulser.

5. Cuvettes are placed on ice for 10 minutes and a 0.1 ml electroshocked sample is added into 0.9 ml TGY recovery medium and incubated anaerobically for 6 hours at 37 °C.

6. After the recovery period, 0.15 ml of TGY culture is plated onto TGY plates containing 50 µg/ml erythromycin and incubated anerobically for 24-48 hours.

▨▨ **Results**

Using the above PEG-based protocol, we were able to transform *C. beijerinckii* NCIMB 8052 with the shuttle vector pMTL500E at a transformation frequency of 7 x10^5 transformants/µg DNA, while PCF112, a shuttle vector based on the indigenous pDM11 plasmid derived from *C. acetobutylicum* NCIMB 6443, was able to transform *C. beijerinckii* at a transformation efficiency of 4 x 10^3 transformants/µg DNA.

Troubleshooting

Because of the sensitivity of *C. beijerinckii* to high applied voltages, it is important not to exceed 10 Kv/cm under the conditions employed. In order to determine the cell lethality following electro-shocking, it is important to plate the post-shock cells on non-selective medium. Our experience suggests that a 3-4 order of magnitude decrease in cell viability allows for optimal transformation efficiency. Also, it is very important to utilize nanopure water (18 m Ω resistance) when making up the electroporation buffer. Using nanopure water plus 1 μg added DNA, it is expected that a pulse duration of the delivered shock would be in the 10-20 msec range. It is also important to minimize exposure of the cells to oxygen during manipulation. Cells should be completely, although gently, suspended in the cuvette and there should be minimal mixing in order to avoid addition of oxygen.

Acknowledgements. Authors would like to thank Satish Herekar for supplying the Bridge Muffy Electroporator. This work was supported in part by the Illinois Corn Marketing Board and the USDA National Research Initiative Biofuels Program.

References

Blaschek HP (1999) *Clostridium* In: Robinson R, Batt C, Patel P (Eds.) Encyclopedia of Food Microbiology. Academic Press, London UK, pp 428-432.

Blaschek HP, White BA (1995) Genetic systems development in the clostridia. FEMS Microbiol Rev 17:349-356

Chen C K, Boucle CM, Blaschek HP (1996) Factors involved in the transformation of previously non-transformable *Clostridium perfringens* type B. FEMS Microbiol Lett 140:185-191.

Dower WJ, Chassy BM, Trevors JT, and Blaschek HP (1992) Protocols for the transformation of bacteria by electroporation. In: Chang DC, Chassy BM, Saunders JA, Sowers AE (eds). Handbook of Electroporation and Electrofusion. Academic Press, New York, pp 485-500.

Kim AY, Blaschek HP (1989) Construction of a *Escherichia coli-Clostridium perfringens* shuttle vector and plasmid transformation of *Clostridium perfringens*. Appl Environ Microbiol 55:360-365.

▨ Suppliers

Gene Pulser™ with Pulse controller:
BioRad
1414 Harbour Way South
Richmond, California 94804
Tel: 1-800-227-5589

Bridge Muffy Electroporator®:
BridgeTechnology
2083 Old Middlefield Way #204
Mountain View, California 94043
Tel: 1-415-988-1548

Electropora-
tors

Electrotransformation of *Lactococcus lactis*

PASCAL LE BOURGEOIS, PHILIPPE LANGELLA and
PAUL RITZENTHALER

Introduction

The last decade has seen a spectacular increase in genetic technology of *Lactococcus lactis*, the model lactic acid bacterium extensively used as starter culture in the manufacture of dairy products. The development of transformation techniques (Gasson & Fitzgerald 1994) and construction of powerful plasmids for gene cloning (de Vos & Simons 1994), as well as for general mutagenesis (Biswas et al. 1993), greatly contributed to this evolution. In the past years, electrotransformation has become the widest used method for introducing DNA in *Lactococcus lactis* cells. Based upon the procedure described by Holo & Nes (1988), the one presented in this chapter is routinely used in our two laboratories and yields 1-2 10^7 transformants per μg of DNA with great reproducibility. Optimization of the procedure was achieved with the *Lactococcus lactis* subsp. *cremoris* strain MG1363 (Gasson 1983), one of the most widely used laboratory strains, using pIL253 (Simon & Chopin 1988) as plasmid DNA. Data obtained from the study of several parameters, such as

i. glycine concentration of the culture broth,

ii. mode of cell freezing,

Pascal Le Bourgeois, LMGM du CNRS, 118 route de Narbonne, Toulouse cedex, 31062, France

Philippe Langella, URLGA, INRA, Domaine de Vilvert, Jouy-en-Josas cedex, 78352, France

✉ Paul Ritzenthaler, LMGM du CNRS, 118 route de Narbonne, Toulouse cedex, 31062, France (*phone* +33-561-335-825; *fax* +33-561-335-886; *e-mail* ritzenth@ibcg.biotoul.fr)

iii. duration of phenotypic expression phase,

iv. composition of plate counting medium,

are presented in the result section. However, it is generally known that electrotransformation efficiency can greatly depend on the strain used, and to a lesser extent on the plasmid type (mode of replication, antibiotic resistance gene,...). As a consequence, electrotransformation procedure could need minor modifications when using any new strain.

Outline

Electrotransformation procedure involved preparation of the cells which were harvested in mid-log phase after incubation in a medium containing high concentration of glycine (used to weaken the cell wall) and 0.5 M sucrose (used as osmotic stabilizer). Following washing steps with buffer containing 0.5 M sucrose and 10% glycerol, cells were concentrated in 30% polyethyleneglycol and 10% glycerol prior to freezing. Bacterial transformation was performed by an electrical shock followed by an expression phase in broth containing 0.5 M sucrose (SGM17). Recovery of transformants was achieved by spreading cells on SGM17 agar plates containing appropriate concentration of antibiotic.

Materials

- 250 ml autoclavable centrifuge bottle (PPCO bottle, Nalgene,   **Equipment**
  USA)
- 40 ml autoclavable centrifuge tube (FEP tube, Nalgene, USA)
- Sterile disposable plastic ware (Sterilin)
- Refrigerated centrifuge (J2-HS, Beckman, USA) with JA-14 and JS-13.1 or JA-20 rotors
- Electroporator with pulse controller (Gene Pulser, Bio-Rad, USA)
- 2 mm electroporation cuvettes (CE-0002, Eurogentec, Belgium)
- Dialysis membrane (VSWP01300, Millipore, USA)

Chemicals
- β-glycerophosphate disodium salt pentahydrate (1.04168, Merck, Germany)
- Agar bacteriological (20001-020, Life Technologies, or 0140-01, Difco, USA)
- Bio-polytone (5 344 1, BioMérieux, France)
- Bio-soyase (5 340 1, BioMérieux, France)
- D(+)-glucose anhydrous (8337, Merck, Germany)
- Glycerol (1.04094, Merck, Germany)
- Glycine (G-7126, Sigma, USA)
- L(+)-ascorbic acid (127, Merck, Germany)
- Magnesium sulfate heptahydrate (5886, Merck, Germany)
- Meat extract (64355, Diagnostic Pasteur, France)
- M17 broth (15029, Merck, Germany)
- Polyethyleneglycol 3000 (819015, Merck, Germany)
- Soluble starch (0178-17-7, Difco, USA)
- Sucrose (1.07651, Merck, Germany)
- Ultrapure water (MilliQ, Millipore, USA)
- Yeast extract (0127, Difco, USA)

Stock broth and solutions
- Commercial two-fold concentrated M17 supplemented with D(+)-glucose (2xGM17): M17 powder 85 g/l and D(+)-glucose 10 g/l in bidistilled water. Autoclaved at 110°C for 30 min and stored at room temperature.
- Home made two-fold concentrated GM17 broth (2xGM17): bio-polytone 10 g/l, bio-soyase 10 g/l, yeast extract 5 g/l, meat extract 8 g/l, L(+)-ascorbic acid 1 g/l, β-glycerophosphate 38 g/l, magnesium sulfate 0.5 g/l, and D(+)-glucose 10 g/l in bidistilled water. Autoclaved at 110°C for 30 min and stored at room temperature.
- Sucrose 1.5 M: 513.75 g/l sucrose in ultrapure water. Autoclaved at 110°C for 20 min and stored at room temperature.
- Glycine 20% (w/v) in bidistilled water. Autoclaved at 110°C for 20 min and stored at room temperature.

Culture broth
- SGM17: 1xGM17 and sucrose 0.5M in sterile bidistilled water.
- SGM17-Gly: 1xGM17, sucrose 0.5M and appropriate concentration of glycine (2.5% for MG1363 strain, 2% for IL1403 strain) in sterile bidistilled water.
- SGM17 plate: SGM17 broth containing 15 g/l agar bacteriological

Sucrose 0.5 M (171.15 g/l), glycerol 10% (w/v) in bidistilled water. Aliquote by 250 ml and sterilize at 110°C for 20 min and store at 4°C.

Washing solution (WS)

Polyethyleneglycol 30% (w/v), glycerol 10% (w/v) in bidistilled water. Aliquote by 5 ml and sterilize at 110°C for 20 min and store at 4°C.

Electrotransformation solution (ES)

As the DNA-cell mixture must have the lowest ionic strength, plasmid DNA or ligation products should be dissolved in bidistilled water. This can easily be achieved either by classical ethanol precipitation, by using QIAquick PCR Purification kit (QIAGEN, Germany), or by membrane dialysis for 15 min against bidistilled water.

Salt removing of ligation mixture prior to electrotransformation

Procedure

All steps in this procedure should be carried out aseptically.

Frozen stock of bacteria pre-adapted in SGM17

Inoculate 5 ml of SGM17 broth by scraping the surface of frozen bacteria and incubate overnight at 30°C. Add 15% (v/v) of sterile glycerol. Vortex the culture to ensure that the glycerol is completely dispersed. Dispense aliquots into sterile storage tubes, and transfer the tubes to -70°C for long-term storage.

Preparation of "electrocompetent" bacteria

1. Inoculate 5 ml of SGM17-Gly broth with frozen bacteria described above and incubate overnight at 30°C.

2. Inoculate 200 ml of SGM17-Gly broth with 100 μl (0.05%) of the overnight culture. Grow the cells at 30°C to an OD_{600} between 0.5 and 0.8.

3. Transfer the culture into a sterile 250 ml centrifuge bottle. Chill the culture immediately on ice.

Note: All subsequent steps should be carried out on ice with pre-cold recipient, rotors and solutions.

4. Centrifuge the cells at 8000 rpm for 10 min at 4°C in a JA-14 rotor (or its equivalent). Discard the supernatant. Stand the bottle in an inverted position to remove traces of medium.

5. Resuspend the pellet, by gentle vortexing or pipetting, in 100 ml (1/2 volume) of WS solution. Centrifuge at 8000 rpm for 10 min at 4°C. Repeat this step once (two washes in WS solution).

6. Resuspend the pellet, by gentle vortexing or pipetting, in 20 ml (1/10 volume) of WS solution. Transfer the cell suspension into sterile 40 ml centrifuge tube. Centrifuge at 6000 rpm for 12 min at 4°C. Discard the supernatant. Stand the bottle in an inverted position to remove traces of washing solution.

7. Resuspend the pellet in 0.8 ml (1/250 volume) of ES solution. Dispense 100 μl aliquots of the suspension into chilled, sterile 1.5 ml Polypropylene tubes (Eppendorf).

8. Freeze the cells by immersing the tubes in liquid nitrogen for 1 min. Store at -70°C until needed.

Transformation of the bacteria

1. Remove a tube of "electrocompetent" cells (one tube per transformation) from the -70°C freezer and thaw on ice.

2. Add a maximum of 10 μl (1/10 volume) of DNA to the cells and incubate on ice for 1 to 5 min. Transfer the mix into ice-cold electroporation cuvettes. Electroporate the cells at 2.5 kV / 25 μF / 400 Ω. This should result in a pulse of 12.5 kV/cm with a time constant (corresponding to the pulse length) between 8 and 9.5 ms.

Note: Arcing can occur i) with old cuvettes and cuvettes that were not cleaned properly, ii) with excess ionic strength of the DNA-cell mixture (see Materials), and iii) if suspension is not in contact with both electrodes of the cuvettes (e.g. presence of air bubbles).

Note: electroporation cuvettes can be reused more than five times without decreasing the transformation efficiency. Cuvettes were sterilized by three washes in distilled water followed by three washes in 70% ethanol, and dried in an inverted position.

3. Immediately add 900 µl of SGM17 broth and homogenize by gentle pipetting. Transfer the discharged cell suspension in a sterile 1.5 ml polypropylene tube and incubate 3 hours at 30°C in a water bath.

4. Spread 0.1 ml of the suspension on SGM17 plates containing the appropriate antibiotic, invert the plates and incubate at 30°C for 24-48 hours. If more than 0.1 ml have to be plated, it was found more convenient to use the pour plate technique (i.e. mixing bacterial suspension with melted agar medium) instead of spreading cells concentrated by centrifugation.

Results

This section provides experimental data obtained after optimization of the original Holo & Nes electrotransformation protocol. Except when noted, MG1363 strain was transformed with 50 ng of CsCl-purified pIL253. Following electrical pulse and expression phase, cells were diluted in SGM17 broth. Fifty microliters of the 10^{-1} or 10^{-2} dilution were spread on solid media supplemented with erythromycin (5 µg/ml) for counting the number of transformed cells, whereas fifty microliters of the 10^{-5} or 10^{-6} dilution were spread on solid media without antibiotic for counting the number of viable cells. This procedure allowed the calculation of both transformation efficiency (number of transformants per µg of supercoiled plasmid) and transformation frequency (number of transformants per surviving cell).

Culture broth origin and freezing conditions

Using MG1363 strain, a slightly higher transformation efficiency was obtained when using home-made GM17 broth. More surprisingly, liquid nitrogen freezing greatly increased the transformation efficiency even when compared with fresh "electrocompetent" bacteria (Fig. 1, front line). To illustrate the need for pro-

CFU / µg pIL253

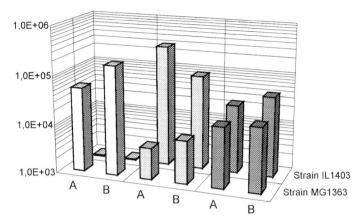

Fig. 1. Effect of the nature of culture broth and freezing conditions on the transformation efficiency. Cells were 100x concentrated in electroporation solution and either frozen with liquid nitrogen (white), or frozen without nitrogen (bright grey) prior transformation, or directly transformed without freezing (dark grey). 100 µl of competent cells were transformed at 2.5 kV / 25 µF / 200 Ω, and incubated for 2 h in SGM17. A: cells grown in commercial broth; B: cells grown in home-made broth. For MG1363 strain, each value was calculated from seven independent experiments (three for IL1403 strain).

tocol optimization when transforming another strain, the same experiments were done on *Lactococcus* lactis subsp. *lactis* IL1403 strain (Chopin et al., 1984). In that case, highest transformation efficiency was obtained on bacteria grown in commercial GM17 broth, and frozen without liquid nitrogen (Fig. 1, back line).

Glycine concentration and pulse length

Holo & Nes (1988) suggested that the optimal glycine concentration to obtain competent cells was usually the highest concentration that allowed growth of the strain of interest. Three glycine concentrations were tested at different pulse lengths (controlled by the resistance values of the pulse controller) for the strain MG1363.

A single electrical pulse was applied using the following parameters: 12.5 kV/cm field strength, 25 µF capacitance, and 200-400 Ω resistances (respectively corresponding to pulse length of 4.6 and 9 ms). Effect of glycine concentration was enhanced

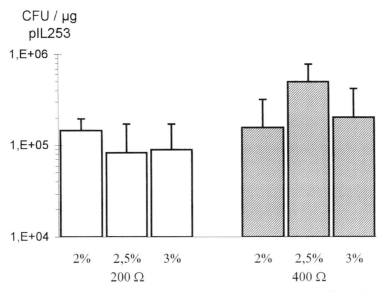

Fig. 2. Effect of glycine concentration and pulse length on the transformation efficiency. Transformation conditions were the same as described in Fig. 1. Each value was calculated from three to seven independent experiments. Error bar is equal to standard error σ_{n-1}.

when electrotransformation was carried out at 400Ω and highest transformation efficiency was obtained when MG1363 bacteria were grown in home-made broth with 2.5 % of glycine (Fig. 2).

Duration of expression phase

Cell viability and transformation efficiency greatly depended on the duration of incubation in SGM17 broth after the electrical pulse (Fig. 3). The highest transformation efficiency was obtained with the greatest reproducibility when incubating the discharged cells for 3 hours. In contrast to the general belief that cell concentration does not significantly increase during expression phase, we observed an increase of viable cells, particularly when adding 950 µl of SGM17 to 50 µl of discharged cells. It cannot be excluded that this increase is due to cell division. If this is true, cell growth could have a detrimental effect on the representativeness of transformants in shotgun cloning strategies (i.e. genomic library). A weaker growth was observed when transforming 100

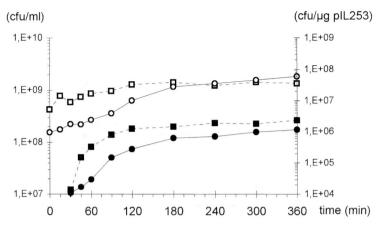

Fig. 3. Effect of duration of expression phase on the number of viable and transformed cells. 100 µl of competent cells (100x concentrated) were transformed at 2.5 kV / 25 µF / 400 Ω, 900 µl of SGM17 was added and aliquots were taken at different times after the electrical pulse. Values were calculated from three independent experiments. Symbols: circles, 50 µl of transformed cells + 950 µl SGM17 (○, number of viable cells; ●, transformation efficiency); squares, 100 µl of transformed cells + 900 µl SGM17 (□ number of viable cells; ■, transformation efficiency).

µl of cells and adding 900 µl of SGM17, and a constant cell concentration of 1-5 10^9 CFU/ml was observed when concentrating the bacteria 1/250 in PG solution (data not shown).

Composition of selective medium

Fig. 4 shows the effect of adding 0.5 M sucrose or 10 g/l soluble starch on GM17 agar broth. Soluble starch increased the transformation efficiency by a factor of 2.5, but had no effect when sucrose was also added to the selective medium. Highest transformation efficiency was obtained when adding 0.5 M sucrose to the selective medium. Adding sucrose or starch on selective plates did not change the viability of the cells but rather increased the transformation frequency (data not shown).

Acknowledgements. We thank Dr. S.D. Ehrlich, E. Maguin, J.C. Piard and Y. Leloir for helpfull discussions. This work was supported by the Centre National de la Recherche Scientifique (UPR9007), and the region Midi-Pyrénées.

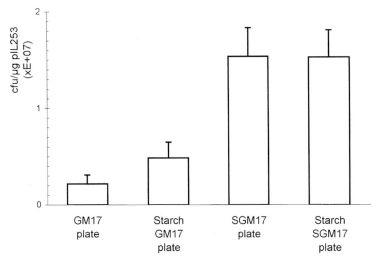

Fig. 4. Effect of soluble starch and/or sucrose in selective plates on the transformation efficiency. 100 μl of competent cells (250x concentrated) were transformed at 2.5 kV / 25 μF / 400 Ω, and incubated for 3 h in SGM17. Each value was calculated from four independent experiments. Error bar is equal to standard error σ_{n-1}.

References

Biswas I, Gruss A, Maguin E (1993) High-efficiency gene inactivation and replacement system for Gram-positive bacteria. J Bacteriol 175:3628-3635

Chopin A, Chopin MC, Moillo-Batt A, Langella P (1984) Two plasmid-determined restriction and modification systems in *Streptococcus lactis*. Plasmid 11:260-263

de Vos WM, Simons G (1994) Gene cloning and expression systems in Lactococci. In: Gasson MJ, de Vos WM (eds) Genetics and biotechnology of lactic acid bacteria. Blackie academic & professional, Glasgow, UK, pp 52-105

Gasson MJ (1983) Plasmid complements of *Streptococcus lactis* NCDO 712 and other lactic streptococci after protoplast-induced curing. J Bacteriol 154:1-9

Gasson MJ, Fitzgerald GF (1994) Gene transfer systems and transposition. In: Gasson MJ, de Vos WM (eds) Genetics and Biotechnology of lactic acid bacteria. Blackie academic & professional, Glasgow, UK, pp 1-51

Holo H, Nes IF (1988) High-frequency transformation, by electroporation, of *Lactococcus lactis* subsp. *cremoris* grown with glycine in osmotically stabilized media. Appl Environ Microbiol 55:3119-3123

Simon D, Chopin A (1988) Construction of a vector plasmid family and its use for molecular cloning in *Streptococcus lactis*. Biochimie 70:559-566

Electrotransformation
of *Salmonella typhimurium*

Natalie Eynard

Introduction

No natural competence is known in Salmonella. Results in the 70's showed that Ca^{2+} shock method was able to give transformation but at very low frequency. With this method, a clear cut dependence on the cell envelope composition was observed. These two draw backs are not present when bacteria are electrotransformed. Electrotransformation appears as a quick and easy to use procedure.

Materials

Strains The classical electrotransformable strains are LT2 and derivatives as SL 1027, CL 4419 and LB 5010 (Binotto, et al., 1991), SL 1306, LB 5000 (O'Callaghan and Charbit, 1990).

Culture media
- LB medium:
  - To 950 ml of deionized H_2O add:
  - 5 g yeast extract (Difco, Detroit, MI)
  - 10 g bactotryptone (Difco)
  - 5 g/l NaCl (Sigma, St Louis, MO).
  - Adjust to 1 liter.
  - Adjust to pH 7.2 with 1N NaOH.
  - Autoclave to sterilize.
- SOC
  - To 950 ml of deionized H_2O add:

Natalie Eynard, IPBS CNRS (UPR 9062), 118 Route de Narbonne, Toulouse, 31062, France (*phone* +33-561-33-58-80; *fax* +33-561-33-58-60; *e-mail* eynard@ipbs.fr)

- 5 g yeast extract (Difco, Detroit, MI)
- 20 g/l bactotryptone (Difco)
- 0.5g NaCl (Sigma)
- 0.2g KCl
- Adjust to 1 liter.
- Adjust to pH 7.2 with 1N NaOH.
- Autoclave to sterilize.
- After cooling at room temperature add:
- 20 ml of a sterile solution of 1 M of glucose (sterilized by autoclaving).
- Just before use add:
- 5 ml of a sterile solution of 2 M of $MgCl_2$ (sterilized by autoclaving)
- 5 ml of a sterile solution of 2 M of $MgSO_4$ (sterilized by autoclaving).
- All liquid media can be stored at room temperature.

Solid media are obtained by addition of agar (15 g/l) before autoclaving. Store at 4°C. For selective media, stock solution of antibiotics is added after autoclaving and cooling at room temperature.

- PB 1   **Pulsing**
    - To 950 ml of deionized H_2O add: **buffer PB**
    - 1 ml of a sterile solution of 1 M of Tris, pH 7.4
    - 270 to 500 ml of 1 M Sucrose (Sigma).
    - Autoclave to sterilize.
    - After cooling at room temperature add:
    - ml of a sterile solution of 1 M of $MgCl_2$ (sterilized by autoclaving).
- PB 2: Sterile water with 10% (w/v) glycerol (Sigma) (sterilized by autoclaving).
- TE buffer
    - To 90 ml of deionized H_2O add:
    - 1 ml of a sterile solution of 1 M of Tris, pH 7.4
    - 20 µl of 0.5M EDTA (Sigma).
    - Adjust to 100 ml.

Procedure

Preparation of "electro-competent" cells

1. A fresh 37°C-overnight culture is diluted from 1:1000 to 1:50 into fresh LB medium.

2. Culture is harvested in the exponential phase (A_{650nm} of 0.4 - 0.6) by centrifugation (4°C, 6000g, 10 min).

3. After removing the medium, the pellet is washed twice in the cold (4°C) pulsing buffer PB or in water and centrifuged again (twice is sufficient in general but some authors make a third wash). With laboratory attenuated strains it is more secure to work under laminar flow hood-(typell).

4. Cells are finally concentrated 100 - 500 fold in cold PB (approximatively 10^{10} C/ml). For the apparatus which delivers square wave pulses, the PB could contain some salts. Presence of magnesium seems to enhance the cell viability of fragile strains and sucrose creates an increase of the external osmolarity which protect the cell during the membrane permeabilization (Eynard, et al., 1992) (PB 1). In the case of a generator producing exponentially decaying pulses, the resistance in the pulsing medium has to be kept as high as possible, and the PB used contains only 10% glycerol (PB 2, (Taketo, 1988)).

5. Cells prepared in this way can be used directly or be frozen (by adding 10% glycerol in the case of the PB 1, or directly with PB 2).

6. Freezing procedure: dispatch 500 µl aliquots in cryo-tubes (Nunc, Kamstrup, DK) and store at - 70°C.

Note: Cells stored in this way can be kept at least one year with no loss in transformability. It is important that all steps from the first centrifugation to the the concentration in PB were made at 4°C.

7. Before use, thaw frozen samples slowly on ice (washed in same volume of the PB when using the PB1).

Preparation of cells-DNA mixture

DNA has to be prepared with minimum level of salt. Purified plasmid can be resuspended in low ionic strength buffer (TE) or simply in water. Because of the restriction barrier between *E. coli* and *S. typhimurium*, the efficiency of transformation is increased about 100 fold when DNA is propagated in *S. typhimurium* (Binotto, et al., 1991).

1. Mix an aliquot of 30 to 40 µl of cells with 1 to 5 µl of DNA. Mixture of cells and plasmid can be done directly in the electropulsation cuvette or in a plastic tube. After a good mix no incubation on ice is needed before applying the pulse. The total volume of the cell DNA mixture depends on the cuvette used for the experiment. But in general an aliquot of 30 to 40 µl of cells and 1 to 5 µl of DNA are used. Approximatively 10^9 cells are electropulsed in an experiment and the transformation efficiency is proportional to the final DNA concentration over a wide range (at least 10 ng/ml to 10 µg/ml, (Binotto, et al., 1991)).

2. Put the mixture in the cuvette
    When filling the cuvette, check that no bubbles are present in the suspension. They may induce arcing when high fields are triggered.

Electropulsation

The electrical parameters depend on the generator used for the experiments.

1. With a square wave pulses delivering generator one pulse of 5 milliseconds duration at 5 kV/cm (Sixou, et al., 1991).

2. With a capacitor discharge generator, apply an exponential decay pulse of 2.5 kV, R= 200 - 400 Ω (a pulse controller is needed), C = 25 µF. The time constant obtained will be about 3 - 5 ms (R = 200 Ω) or 5 -10 ms (R = 400 Ω), (Binotto, et al., 1991) and the initial field strength reaches 12.5 kV/cm for a 0.2 cm cuvette.

The initial value of the applied field is always higher with a capacitor discharge system than the mean value set for a square wave generator. For exponential decaying pulse, the field applied is not over the critical value needed for transformation during all the pulse. Salts can be present when using a square wave pulse because they do not affect the pulse duration but a high salt concentration should be avoided because it causes a strong temperature increase.

Expression of transformants

1. Immediately after pulsing, the electropulsed volume is mixed with 1 ml of culture medium at room temperature. Mix by shaking the cuvette. This is very important when using the 0.1 cm cuvette to obtain an homogeneous suspension.
    LB can be used but higher efficiencies are obtained with SOC medium (O'Callaghan and Charbit, 1990).

2. Then the mixture is put on sterile tube and incubated for 1 h at 37°C (shaking is not necessary).

3. The cells are then plated on solid medium containing the appropriate selecting agent and incubated at 37°C.
    The number of transformants depends on the quantity of DNA put in the electropulsation cuvette, the dilution used and the volume of aliquot spread on the plate (classically 100µl, but if the dish is lacking in moisture, it can be filled with 300 or 400 µl when expected efficiency is low).
    The efficiency E = transformants on the plate x dilution x (10^{-3} / volume spread on the plate) x (10^{-6} / quantity of DNA in the cuvette). 10^{-3} is for the total volume of electropulsed cells (approximatively 1 ml with the culture medium) and 10^{-6} represents the 1 µg DNA used for expression of efficiency.

4. Colony forming units (CFU) can be counted after an overnight or 24 H incubation.

5. To estimate the cell survival, an 100 µl aliquot can be plated on non selective medium with a preliminary dilution 10^{-5} or 10^{-6} (if 10^9 cells were pulsed).

References

Binotto, J., P. R. MacLachlan and K. E. Sanderson. 1991. Electrotransformation in Salmonella typhimurium LT2. Can J Microbiol. 37:474-477.

Eynard, N., S. Sixou, N. Duran and J. Teissie. 1992. Fast kinetics studies of Escherichia coli electrotransformation. Eur J Biochem. 209:431-436.

O'Callaghan, D. and A. Charbit. 1990. High efficiency transformation of Salmonella typhimurium and Salmonella typhi by electroporation. Mol Gen Genet. 223:156-158.

Sixou, S., N. Eynard, J. M. Escoubas, E. Werner and J. Teissie. 1991. Optimized conditions for electrotransformation of bacteria are related to the extent of electropermeabilization. Biochim Biophys Acta. 1088:135-138.

Taketo, A. 1988. DNA transfection of Escherichia coli by electroporation. Biochim Biophys Acta. 949:318-324.

Electroporation of bifidobacteria

MADDALENA ROSSI, PATRIZIA BRIGIDI and DIEGO MATTEUZZI

Introduction

Bifidobacterium represents one of the most numerous bacterial genera in the gut of humans and other animals and plays a fundamental role in the health of the host. In view of the development of bifidobacteria probiotic strains with improved characteristics, there is considerable interest in their manipulation using genetic engineering techniques. In spite of their importance, very little is known about the genetics of these microorganisms. Application of recombinant DNA technology to bifidobacteria was delayed by past difficulties associated with the transformation of members of this genus. Only recently has the successful transformation of *Bifidobacterium* by electroporation been reported (Missich et al. 1994; Argnani et al. 1996, Rossi et al. 1997; Matsumura et al. 1997). An optimized procedure for the electroporation of plasmid DNA into several *Bifidobacterium* species is described.

Maddalena Rossi, University of Bologna, Department of Pharmaceutical Sciences, Interdepartmental Center for Biotechnology, Via Belmeloro 6, Bologna, 40126, Italy
Patrizia Brigidi, University of Bologna, Department of Pharmaceutical Sciences, Interdepartmental Center for Biotechnology, Via Belmeloro 6, Bologna, 40126, Italy
✉ Diego Matteuzzi, University of Bologna, Department of Pharmaceutical Sciences, Interdepartmental Center for Biotechnology, Via Belmeloro 6, Bologna, 40126, Italy (*phone* +39-51-2099733; *fax* +39-51-2099734; *e-mail* matt@alma.unibo.it)

Materials

Bifidobacteria were grown anaerobically at 37°C in IM broth: **Media and buffers**

Bacto-tryptone (Difco Laboratories)	10 g
Bacto-yeast extract (Difco Laboratories)	5 g
Bacto-tryptose (Difco Laboratories)	5 g
Lactose	3 g
Di-ammonium hydrogen citrate	2.5 g
Sodium pyruvate	1.2 g
Cysteine-HCl	0.3 g
Tween 80	1 g
MgSO4 · 7 H2O	0.5 g
MnSO4 · 4 H2O	0.12 g
FeSO4 · 7 H2O	0.03 g
Distilled water	1 l

pH adjusted to 6.8 with NaOH.

Sterilized at 110°C for 30 min

– KMR buffer: KH_2PO_4 5 mM, $MgCl_2$ 1 mM, raffinose 0.3 M, pH 4.8:
  5.35 g of raffinose were dissolved into 30 ml of KH_2PO_4 5 mM, then 15 µl of $MgCl_2$ 2 M were added. The solution was sterilized using a 0.45 (m filtration unit. Before use, the buffer was cooled on ice for an hour. Stable for up to 3 months at room temperature.
  Anaerobic conditions were obtained by BBL Gas Pack System, Becton Dickinson, USA.

The experiments were carried out with a Bio-Rad Gene Pulser™   **Equipment**
apparatus (Bio-Rad Laboratories, USA) equipped with a Pulse Controller and 0.2-cm gap disposable cuvettes (Bio-Rad).

Procedure

Preparation of electrocompetent cells

1. Inoculate 120 ml of IMA broth with 6 ml of a stationary phase (16-18 h) culture grown in the same medium. After the inoculum, the O.D.$_{600}$ of the culture should be 0.09-0.12.

2. Grow anaerobically at 37°C.

3. At the early-log phase (O.D.$_{600}$ 0.2-0.3), approximately after 3-4 h incubation, chill the culture on ice for 10 min.

4. Harvest the cells in a cold rotor at 5000 × g for 10 min.

5. Resuspend the pellet in 60 ml of cold K-phosphate buffer 5mM pH 7 and centrifuge as above.

6. Resuspend the pellet in 360 µl of cold KMR buffer.

7. Store the cell suspension overnight on ice.

Electroporation

1. Into a cold polypropylene tube, mix 80 µl of cell suspension with 0.25 µg of plasmid DNA dissolved in 1-5 µl of TE. Store the sample on ice for 5 min.

2. Transfer the suspension to a cold 0.2-cm sterile electroporation cuvette and pulse at 12.5 kV, 25 µF, 100 Ω. The pulse duration should range from 2.3 to 2.4 msec.

3. Immediately add 1 ml of IMA broth at room temperature to the cuvette and recover the cells, with gentle mixing, into a culture tube.

4. Incubate the sample anaerobically at 37°C for 3 h.

5. Plate directly on selective IMA plates.

Results

Bifidobacterium animalis MB209 (=ATCC 27536) was used to develop the method described and to evaluate the influence of several variables on transformation efficiency. Therefore, several wild type strains of *Bifidobacterium* were transformed with pNC7 (Rossi et al. 1996) applying the optimal conditions determined for *B. animalis* MB209. As shown in Table 1, electroporation efficiencies ranged from 3.6×10^1 to 1.2×10^5. These efficiencies were obtained with plasmid pNC7 purified from *B. animalis* MB209, to avoid any restriction barriers. pNC7 is a recombinant plasmid that replicates only in bifidobacteria, based on the native plasmid pMB1 isolated from *B. longum* B2577 (Matteuzzi et al. 1990). The wide range of electroporation efficiencies suggested that the strain of recipient bacterium plays a fundamental role in the transformation yields. The source of plasmid had great effect on transformation efficiency of bifidobacteria. When *B. animalis* MB209 was electroporated with the shuttle vectors *E. coli-Bifidobacterium* pDG7 or pLF5 (Matteuzzi et al. 1990; Rossi et al. 1998) isolated from *Bifidobacterium* spp., efficiencies were similar to that obtained by using pNC7. However, pDG7 or pLF5 isolated from *E. coli* HB101 reduced the efficiency by a factor of 30. The presence of 16% Actilight®P into IM broth played an important role in the process of DNA uptake: in cells grown in IMA a 100-fold efficiency increase was observed, in respect to cells grown in IM. High electroporation efficiencies were obtained also electroporating cells grown in IM broth supplemented with raffinose 0.3 M (Rossi et al. 1997). However, the different transfer efficiency was not induced by the faster growth rate of the culture, because all the tested strains grew slower in presence of 16% Actilight or raffinose 0.3 M than in IM broth. Presumably the presence of Actilight®P or raffinose affected the cell wall thickness, density and structure. Furthermore, transformation efficiencies were optimal with cells harvested at the early exponential phase, while dropped off for older cells, until they reached zero for cells from stationary phase. Although the procedure recommended the setting of 25 µF, 100 Ω and 12.5 kV/cm, the highest transformation rates were obtained when cells were exposed to a single pulse of 25 µF, 200 Ω and 12.5 kV/cm. However at the resistance of 200 Ω, arcing occasionally occurred.

Table 1. Transformation efficiencies of *Bifidobacterium* strains (collection of the Institute of Agricultural Microbiology, University of Bologna, Italy) belonging to 13 different species (KMR buffer, 12.5 kV/cm, 100 Ω, 25 µF)

Strains	Transformants per mg of pNC7 DNA
B. longum MB219	4.1×10^2
B. longum MB260	9.3×10^2
B. bifidum MB254	7.2×10^4
B. breve MB226	6.6×10^4
B. breve MB252	2.3×10^4
B. infantis MB208	1.2×10^5
B. infantis MB263	9.3×10^3
B. infantis-longum MB231	2.8×10^2
B. animalis MB209 (=ATCC27536)	3.0×10^4
B. pseudocatenulatum MB264	5.0×10^1
B. ruminale MB266	7.2×10^2
B. dentium MB269	3.6×10^1
B. magnum MB267	1.8×10^3
B. cuniculi MB279	1.5×10^2
B. cuniculi MB280	8.1×10^2
B. cuniculi MB281	7.5×10^3
B. indicum MB101	7.8×10^2
B. asteroides MB100	8.9×10^2

References

Argnani A, Leer RJ, van Luijk N, Pouwels PH (1996) A convenient and re-producible method to genetically transform bacteria of the genus *Bifidobacterium*. Microbiology 142: 109-114

Matsumura H, Takeuchi A, Kano Y (1997) Construction of *Escherichia coli-Bifidobacterium longum* shuttle vector transforming *B. longum* 105-A and 108-A. Biosci Biotech Biochem 61: 1211-1212

Matteuzzi D, Brigidi P, Rossi M, Di Gioia D (1990) Characterization and molecular cloning of *Bifidobacterium longum* cryptic plasmid pMB1. Lett Appl Microbiol 11: 220-223

Missich R, Sgorbati B, LeBlanc DJ (1994) Transformation of *Bifidobacter-ium longum* with pRM2, a constructed *Escherichia coli-B. longum* shuttle vector. Plasmid 32: 208-211

Rossi M, Brigidi P, Gonzalez Vara A, Matteuzzi D (1996a) Characterization of the plasmid pMB1 from *Bifidobacterium longum* and its use for shuttle vector construction. Res Microbiol 147: 133-143

Rossi M, Brigidi P, Matteuzzi D (1996b) An efficient transformation system for *Bifidobacterium* spp. Lett Appl Microbiol 24: 33-36

Rossi M, Brigidi P, Matteuzzi D (1998) Improved cloning vectors for *Bifidobacterium* spp. Lett Appl Microbiol

Suppliers

DIFCO LABORATORIES
PO Box 331058
Detroit MI 48232-7058
USA
tel.: +01-313-462-8500 or -800-521-0851
fax: +313-462-8517

BEGHIN-MEIJI INDUSTRIES
54, Avenue Hoche
75360 PARIS CEDEX 08
FRANCE
tel.: +33-1-40535786
fax: +33-1-47637447
telex 644352

BECTON DICKINSON
Becton Dickinson Microbiology System
PO Box 243
Cockeysville MD 21030
USA
tel.: +01-800-638-8663

BIO-RAD Laboratories
200 Alfred Nobel drive
Hercules CA 49547
USA
tel.: +01-510-741-1000 or -1-800-424-6723
fax: +01-510-741-5800 or -1-800-879-2289
telex 335-358

Electrotransformation of Listeria species

JANET E. ALEXANDER

Introduction

Electroporation offers an efficient system for the introduction of genetic material into listeriae. Conjugation and protoplast transformation of *Listeria monocytogenes* has been reported (Flamm et al. 1984, Vicente et al. 1987). However, transformation rates ($\sim 10^3/$ µg of DNA) produced by these methods are too low for use in many molecular techniques. To facilitate the entry of DNA by electroporation, sufficient intensity of current must be applied to produce the optimum number of permeablilized areas. The intensity of current is a function of the field strength and time constant used.

- The field strength is defined as the voltage gradient between the electrodes.

- The time constant is a function of the field strength and the resistance applied.

- The resistance applied is determined by the parallel connection of different resistors (Bio-Rad Manual, 1994).

Janet E. Alexander, University of Leicester, Department of Microbiology and Immunology, Medical Sciences Building, University Road, Leicester, LE1 9HN, United Kingdom (*phone* +44-(0)116-2523018; *fax* +44-(0)116-2525030; *e-mail* JEC@LEICESTER.ac.UK)

▨ Outline

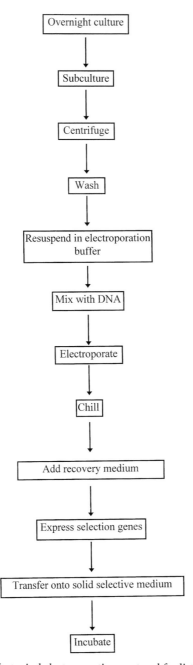

Fig. 1. The steps of a typical electroporation protocol for listeriae are outlined

▓ Materials

Equipment – Gene-Pulser^{TM} electroporation apparatus and pulse controller (Bio-Rad, Hercules, California, USA). A pulse controller is essential for safe use of electroporator.
  – Electroporation cuvettes (0.2 cm^2) (Bio-Rad, Hercules, California, USA)

Electroporation buffers

Sucrose magnesium electroporation medium (SMEM) (pH 7.2)
– 9.2M sucrose
– 3.5mM MgCl$_2$
  Sucrose should be filter sterilised and added to autoclaved MgCl$_2$ solution. Bring buffer to pH 7.2 with the minimum of glacial acetic acid and 0.1 M NaOH. Store at 4°C for up to 1 month.

HEPES-sucrose electroporation medium (HEPES-S) (pH 7.0)
– 1mM HEPES
– 0.5M sucrose
  Sucrose should be filter sterilised and added to autoclaved MgCl$_2$ solution. Bring buffer to pH 7.0 with the minimum of glacial acetic acid and 0.1 M NaOH. Store at 4°C for up to 1 month.

Media – Brain Heart Infusion Broth (BHI) (Oxoid)
  – Blood Agar Base No.2 (BAB) (Difco)

Subprotocol 1
Preparation of Competent Cells for the Electroporation of Listeria Species

▓▓ Procedure

Cell preparation

1. Add a 10% (v/v) inoculum of an overnight culture of Listeria in BHI to 500 ml of BHI.

2. Incubate with shaking (~150 rpm) at 37°C to an OD$_{600}$ of 0.6-0.8.

3. Harvest bacteria by centrifugation at 5000 x g for 10 min at 4°C.

4. Wash twice in 1/10 th the original volume with ice cold SMEM.

5. Resuspended in 1/100 th the original volume with ice cold SMEM.

6. Portions of the cell suspension are stored at -70°C for up to one month.

Electroporation of Listeria species with plasmid DNA

1. Thaw prepared cells slowly on ice.

2. Make initial viable colony forming unit (cfu) count for calculation of % survival.

3. Mix 40 µl (~4 x 10^{10} bacteria) gently with 1µl of DNA (~0.5 µg) dissolved in sterile nanopure distilled water.

4. Leave on ice for 1 min.

5. Transfer the mixture to a chilled (on ice for at least 15 mins) electroporation cuvette.

6. Place between the chilled electrodes of a Gene-Pulser^{TM} electroporation apparatus.

7. Electroporate at a field strength of 8.5 kV/ cm, 200 Ohms resistance 25 µF capacitor with a time constant of approximately 5.0 ms.

8. Immediately place the cuvette on ice for 1 min.

9. Add 1 ml of BHI broth and make final viable cfu count for calculation of % survival.

10. Incubate at 37°C for 2 hrs with gentle shaking (~100 rpm).

11. Plate serial dilutions of the cells onto BAB containing selective levels of antibiotic.

12. Incubate at 30°C for 48 hrs.

Electroporation

▨▨ Results

Electroporation of Listeria monocytogenes

The results of a typical experiment to determine the optimal conditions for electroporation of *L. monocytogenes* NCTC 7973 with pGK12 plasmid DNA (Kok et al. 1984) are shown in Table 1. The procedure was conducted at various field strengths for different time constants. The time constant is a function of the field strength and the resistance applied (Bio-Rad Manual, 1994). A range of time constants at each field strength studied was achieved by the connection of different resistors in parallel with the sample.

Table 1. Electroporation of *Listeria monocytogenes* NCTC 7973 with pGK12 DNA

Field strength (kV/cm)	Time constant (ms)	Actual time constant[a] (ms)	Percentage survival	No. of transformants (µmg DNA)
3.1	5.0	4.4	57.3	0.0
3.1	10.0	11.3	51.4	3.0 x 102
3.1	15.0	14.6	42.6	2.9 x 102
3.1	20.0	16.8	9.00	2.5 x 102
6.2	5.0	4.4	33.6	3.3 x 104
6.2	10.0	11.7	35.5	7.3 x 104
6.2	15.0	13.8	0.4	3.5 x 103
6.2	20.0	Arc[b]	–	–
8.5	2.0	2.3	89.8	2.4 x 104
8.5	4.0	4.3	59.9	1.3 x 105
8.5	5.0	4.4	61.9	3.9 x 106
8.5	8.0	6.8	24.8	1.1 x 104
8.5	10.0	Arc	–	–
8.5	15.0	Arc	–	–
8.5	20.0	Arc	–	–

[a] Actual time constants are given due to small variations in conductivity of samples and cuvettes.

[b] Arc indicates pulse did not pass through sample.

The results of the experiments in Table 1 indicate that both the field strength and time constant influence the degree of transformation but for each field strength there was an optimal time constant. The application of a field strength above 8.5 kV/cm exceeded the conductivity of the electroporation medium and resulted in arcing. Up to 8.5 kV/cm an increase in field strength favoured successful transformation at each time constant. The efficiency of transformation also increased with the time constant at each of the field strengths used up to a certain time constant after which efficiency declined, possibly due to the increased killing of the bacteria. If a short time constant (~2 ms), was used to reduce cell death, at field strengths of 3.1 or 6.2 kV/cm no transformants were recovered; transformation was maximal at 10 ms. At 8.5 kV/cm the transformation rate increased with time constant and was maximal at 5 ms. These results are probably due to incomplete permeabilization of the cell membrane at the lower time constant.

The percentage of bacteria killed during electroporation is an important consideration. At any field strength an increase in the time constant results in increased killing (Table 1). This limits the use of longer time constants just as the conductivity limit of the electroporation buffer limits the use of increased field strengths. Therefore the optimum time constant amd field strengths are those that limit the percentage of the bacterial population killed to a minimum, while still allowing maximum polarization of the membrane to allow DNA entry. On the basis of the results in Table 1 the optimum conditions for the recovery of transformants of *Listeria monocytogenes* (NCTC 7973) with plasmid pGK12 were judged to be a field strength of 8.5 kV/cm, 200 Ohms resistance, 25 µF capacitor, with a time constant of ~ 5 ms. Under these conditions, a transformation frequency of approximately 4 x 10^6 / µg pGK12 DNA was achieved.

Troubleshooting

- Arcing (current did not pass through sample)
  - Conductivity limit of the electroporation buffer exceeded, check composition of buffer, reduce field strength.
  - Salt residues present in DNA, dialise DNA, reduce volume of DNA added.

- Low percentage survival
  - Prepared cells stored for too long, at a temperature $> -70°C$, or thawed too quickly not on ice.
  - Time constant too long, check settings.
  - Electroporation cell not sufficiently chilled, leave on ice for at least 15 mins prior to use to minimise heating of sample.
  - Composition of electroporation buffer not correct. Sucrose is an important agent for the survival of the permeabilised cells, the ionic strength of the electroporation medium not only determines the current passing through the sample but also the rate of heating. Higher conductivity is likely to reduce cell survival, due to increased heating of the sample by the greater current passing through it. SMEM also contains $MgCl_2$, its concentration is important because Mg^{+2} has been shown to reduce cell kill by stabilising the cell wall (Miller, 1988).

- Low frequency of transformation
  - Arcing, time constant too short or field strength too high; check settings. Frequency of transformation may be improved by the use of a medium with a higher ionic strength and greater conductivity limit. Improved conductivity of the medium can be achieved by the omission of osmotic agents such as the sucrose in the SMEM. However, this may reduce % survival.
  - Gene expression period too short, increase incubation in BHI post electroporation.
  - Restriction barriers to the entry of DNA prepared in E. coli may be present (Shigekawa & Dower, 1988) this may be present in some spp. The problem may be overcome by replicating the DNA of interest in L. monocytogenes before extraction for electroporation.
  - Large size plasmid DNA to be introduced. Gram-positive bacteria appear to be particularly sensitive to the size of the transforming plasmid DNA. In general, a greater degree of cell wall permeabilisation appears to be required for the transformation of Gram-positive bacteria with plasmid DNA in excess of 10 kb (Trevors, et al. 1992). This problem may be overcome by incorporating penicillin treatment into the electroporation protocol to increase the permeabilisation of the cell wall and thus allow the entry of large vectors.

Subprotocol 2
Electroporation of Listeria Species
with Vectors Greater than 10 kb

The use of penicillin pretreatment of cells for electroporation has been reported to be successful for *L. monocytogenes* by Park & Stewart (1990). Cell wall peptide crosslinking between glycan chains is inhibited by Penicillin G treatment. This destabilises the cell wall and increases the degree of permeabilisation during electroporation to facilitate the entry of large vectors. Unfortunately, during this process the cells become very osmotically fragile because cell wall renewal is retarded, making the recovery of electrotransformants difficult. For this reason sucrose was included to osmotically stabilise the recovery medium, and the incubation period after electroporation was extended from the 2 hrs used for the untreated electrotransformants, to 4 hrs. These steps allow regeneration of a functional cell wall, as well as expression of antibiotic resistance, before plating onto agar.

▓▓ Procedure

1. Add a 20% (v/v) inoculum of an overnight culture of *L. monocytogenes* in BHI 0.5 M sucrose (BHI/s) to 500 ml of fresh BHI/s. **Preparation of cells**

2. Incubate at 37°C with shaking (~150 rpm) to an OD_{600} of 0.2.

3. Add Penicillin G to a final concentration of 10 μg/ml, and continue incubation for a further 2 hrs (OD_{600} 0.35-0.40).

4. Harvest cells by centrifugation at 7,000 x g for 10 min at 4°C.

5. Wash twice in an equal volume of ice cold HEPES-S.

6. Resuspended cell pellet in 1/400th vol. of ice cold HEPES-S.

7. Place cells on ice and use immediately for electroporation.

1. Mix 100 μl of the prepared cell suspension (~1 x 10^{10} bacteria) with 25 μl of DNA (~1μg) dissolved in sterile nanopure distilled water. **Electroporation**

2. Leave on ice for 1 min then transfer to a chilled electroporation cuvette.

3. Place between the chilled electrodes of a Gene-Pulser™ electroporation apparatus.

4. Electroporate cells at a field strength of 10 kV/cm, 200 Ohms resistance, 25 µF capacitor with a time constant of approximately 4.0 ms.

5. Immediately place cuvette on ice for 1 min.

6. Add 1 ml of BHI/s and incubated at 37°C for 4 hrs.

7. Plate cells onto BHI - 1.5% agar, containing selective levels of antibiotic.

8. Incubate plates at 30°C for 48 hrs.

Results

Penicillin treatment before electroporation allowed the successful transformation of *L. monocytogenes* with plasmid pLTV3 (22.1 kb, Camilli et al. 1990) DNA at a frequency of 2.3×10^2 transformants /µg. No transformants were recovered when penicillin treatment was not included.

References

Bio-Rad Manual, (1994). Gene Pulser Transfection Apparatus Operating Instructions and Application Guide. Bio-Rad, Hercules, California, USA.

Camilli, A., Portnoy, D. A., & Youngman, P. (1990). Insertional mutagenesis of *Listeria monocytogenes* with a novel Tn917 derivative that allows direct cloning of DNA flanking transposon insertions. Journal of Bacteriology. 172: 3738-3744.

Flamm, R. K., Hinrichs, D. J., & Thomasow, M. (1984). Introduction of pAMB1 into *Listeria monocytogenes* by conjugation and homology between native *Listeria monocytogenes* plasmids. Infection and Immunity. 44: 157-161.

Kok, J., Van Der Vossen, J. M. B. M., & Venema, G. (1984). Construction of plasmid cloning vectors for lactic streptococci which also replicate in *Bacillus subtilis* and *Escherichia coli*. Applied Environmental Microbiology. 48: 726-731.

Miller, J. F. (1988). Bacterial Electroporation. Molecular Biology Reports. Bio-Rad Laboratories, Bio-Rad, Richmond, California, USA. 5: 1-4.

Park, S. F., & Stewart, G. S. A. B. (1990). High-efficiency transformation of *Listeria monocytogenes* by electroporation of penicillin-treated cells. Gene. 94: 129-132.

Shigekawa, K., & Dower, W. J. (1988). Electroporation of eukaryotes and prokaryotes: a general approach to the introduction of macromolecules into cells. Biotechniques. 6: 742-751.

Trevors, J. T., Chassy, B. M., Dover, W. J., & Blaschnek, H. P. (1992). Electrotransformation of bacteria by Plasmid DNA. In; Guide to Electroporation and Electrofusion. Academic Press Inc, London.

Vicente, M. F., Baquero, F., & Prez-D'az, J. C. 1987. A protoplast transformation system for *Listeria spp*. Plasmid. 18: 89-92.

Suppliers

Bio-Rad Laboratories Ltd, 2000 Alfred Nobel Drive, Hercules, California 94547, USA. Phone: +510-232-7000.

Oxoid Ltd. Wade Rd, Basingstoke, Hampshire RG24 8PW, UK. Phone: +44-(0)1256-841144, fax: +44-(0)1256-463388.

Difco Laboratories. 17197 North Laurel Park Dr. Livonia, Michigan 48152, USA. Phone: +313-462-8500.

Abbreviations

OD_{600}	optical density at 600nm.
NCTC	National Type Culture Collection, London, UK.
ACTC	American Type Culture Collection, Rockville, USA.
CIP	Collection Institut Pasteur, Paris, France.

Transformation of *Methylobacterium extorquens* with a Broad-Host-Range Plasmid by Electroporation

SHUNSAKU UEDA and TSUNEO YAMANE

Introduction

In genetic studies with methylotrophic bacteria as the host organisms, the conjugative transfer system has been widely used for introducing DNA molecules into the cells. Since this system is sometimes technically cumbersome and time-consuming, electroporation technique has now been developed in order to introduce DNA molecules into some types of cells such as mammalian cells, plant protoplasts, yeast cells, and bacterial cells. However, an electric transformation for methylotrophic bacteria has been very seldom reported. Here, the electropolation system is applied to introduce a broad-host-range plasmid pLA2917 into a methylotrophic bacterium *Methylobacterium extorquens*, as a model system.

Materials

Electroporation apparatus A somatic hybridizer model SSH-2 (Shimadzu Co., Kyoto, Japan) was used. The model SSH-2 generates a square-shaped pulse up to 700 V at a duration time of 100 to 500 μsec. The electroporation chamber used was a type SSH-C11 with the electrodes set 0.5 mm apart. Therefore, a maximum field strength of 14 kV/cm is

✉ Shunsaku Ueda, Utsunomiya University, Department of Bioproductive Sciences, Faculty of Agriculture, 350 Mine-machi, Utsunomiya, 321-8505, Japan (*phone* +81-28-649-5475;
fax +81-28-649-5401; *e-mail* uedashun@cc.utsunomiya-u.ac.jp)
Tsuneo Yamane, Nagoya University, Division of Molecular Cell Mechanisms, Department of Biological
Mechanisms and Functions, Graduate School of Bio- and Agro-Sciences, Chikusa-ku, Nagoya, 464-8601, Japan

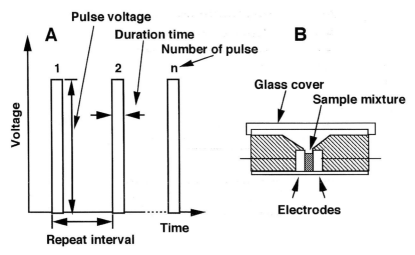

Fig. 1. Electric pulse generated by the electroporation apparatus model SSH-2 (A) and the cross section of the chamber SSH-11.

possible with this chamber. Maximum sample volume of the chamber is 10 μl. The characteristics of the electric pulse generated by the model SSH-2 and the cross section of the chamber are illustrated in Figure 1A and B, respectively.

10mM Tris-HCl, 2mM MgCl₂(6H₂O, 10% (wt/vol) sucrose (pH7.5) | **Electroporation buffer**

A broad-host-range plasmid pLA2917 (21 kb, kanamycin and tetracycline resistances) (Allen et al. 1985) was used. pLA2917 was prepared by a standard alkali-sodium dodecyl sulfate method (Birmbiom et at. 1979) from an *Escherichia coli* HB101 harboring the plasmid. | **Plasmid DNA**

Procedure

1. Grow the cells of *M. extroquens* aerobically to the middle exponential phase (1.4 x 10¹⁰ / ml) in an inorganic salt medium (Sato et al. 1977) containing 1% (vol/vol) methanol [MIS medium] as a sole carbon source at 30 °C.

2. Harvest the cells by centrifugation at 6,000 x g for 10 min.

3. Wash the cells with an appropriate volume of electroporation buffer.

4. Resuspend the cells in the same buffer at a cell concentration of 7 x 10^{10} / ml.

5. Mix the cell suspension and the plasmid solution (70 mg/ml) at a ratio of 9:1 (vol/vol).

6. Transfer 10 µl of the mixture into a space between the electrodes of electroporation chamber sterilized by autoclaving.

7. Subject the mixture to electric pulse (typical electric conditions: field strength; 10kV/cm, pulse duration; 300 µsec; and number of pulse; 10 times; pulse repeat interval, 0.5 sec).

8. Transfer the mixture to a small tube containing 0.2 ml of the MIS medium.

9. Incubate the tube for 2 h at 30°C to allow expression of antibiotic resistance genes of the plasmid.

10. Spread the mixture on an MIS agar plate containing kanamycin (50 µg/ml).

11. Incubate the selective plate at 30 °C for 4 to 5 days and score transformants displaying kanamycin resistance.

Results

The transformation of *M. extorquens* by electroporation was affected by electric conditions such as field strength, pulse duration, and number of pulse. A high efficiency was generally obtained with electric condition that killed around 50% of the viable cells, which is shown in Figure 2 and Table 1, as typical results. The condition was a combination of field strength of 10 kV/cm, pulse duration of 300 µsec, and pulse number of 10. Growth phase of the cells was also an important factor in obtaining a high transformation efficiency. To use the exponentially growing cells was essential, because the transformation efficiency was drastically decreased with the cells at stationary phase. The number of transformants obtained was dependent on DNA concentration under the optimum electric condition (Table 2).

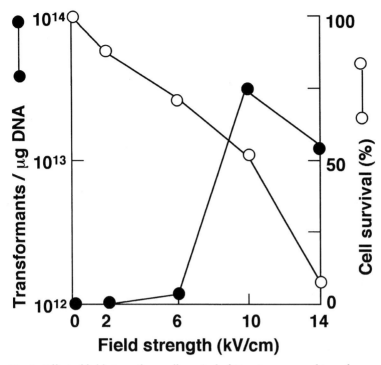

Fig. 2. Effect of field strength on cell survival of *M. extorquens* and transformation efficiency. The duration time and number of pulse were 300 msec and 10, respectively.

Table 1. Effects of electric conditions on transformation efficiency

		Transformants/µg DNA
Exp. 1	Pulse duration (µsec)	
	100	4.4×10^2
	200	1.5×10^3
	300	7.6×10^3
	400	3.8×10^3
	500	2.9×10^3
Exp. 2	Number of pulse	
	1	1.2×10^2
	3	5.5×10^2
	6	2.5×10^3
	10	7.6×10^3
	15	7.0×10^3
	20	7.0×10^3

Exp. 1: Field strength, 10 kV/cm; number of pulse, 10

Exp. 2: Field strength, 10 kV/cm; pulse duration, 300 µsec

Table 2. Effect of DNA concentration on transformation[a]

DNA concentration (mg/ml)[b]	Transformants
0	0
0.1	65
0.5	153
1.0	248
2.5	425
5.0	572
7.5	615

a) Electric conditions: field strength. 10 kV/cm; pulse duration, 300 µsec; number of pulse, 10.

b) 1 µl of each DNA sample was used for electroporation

Applications

Here, we described transformation of *M. extorquens* with pLA2917 by electroporation. In addition to this strain, we tried to transform other methylotrophic bacteria belonging *Methylobacterium* sp. (*M. organophilum, M. mesophilum, M. rhodos, M. radiotolerans, M. zatmanii*) with recombinant plasmids constructed during the study on replication origin region of a cryptic plasmid harbored by *M. extorquens* (Results will be reported elsewhere). Since it was found that electroporation was effective in transforming these bacteria with the recombinant plasmids, the method described here will be applicable in the genetic studies of methylotrophic bacteria.

References

Allen LN, Hanson RS (1985) Construction of broad-host-range cosmid cloning vectors: identification of genes necessary for growth of *Methylobacterium organophilum* on methanol. J Bacteriol 161:955-962.

Birmboim HC, Doly J (1979) A rapid alkaline extraction prodedure for screening recombinant plasmid DNA. Nucreic Acids Res 7:1513-1523.

Sato K, Ueda S, Shimizu S (1977) Form of vitamin B12 and its role in a methanol-utilizing bacterium, *Protaminobacter ruber*. Appl Environ Microbiol 33: 515-521.

Electrotransformation of Acidophilic, Heterotrophic, Gram-negative Bacteria

THOMAS E. WARD

Introduction

Members of the genera *Acidiphilium* and *Acidocella* are obligately acidophilic, heterotrophic, Gram-negative, aerobic, mesophilic eubacteria (Wichlacz and Unz 1981; Harrison 1984, 1989; Kishimoto et al 1995). They inhabit a variety of low pH environments including acidic sulfide ore leaching and coal mine drainage sites (Harrison 1984), generally created by chemoautotrophic bacteria which oxidize reduced sulfur species to sulfuric acid, e.g. *Thiobacillus ferrooxidans* and *T. thiooxidans*. The exact mechanism(s) by which they are adapted to these highly acidic environments have not been completely elucidated. Such bacteria have potential applications in biomining (recovery of metals from ores using microbiological leaching); in the treatment of acidic hazardous waste; in biological removal of sulfur from coal; and indeed in any bioprocessing applications in which it is necessary or advantageous to operate under acidic conditions.

In the course of our research to develop genetic systems for these acidophiles, several methods for introducing isolated DNA into bacteria, including $CaCl_2$ shock, DMSO plus DTT, and production of lysozyme spheroplasts, with and without polyethylene glycol induced fusion, were attempted without success (Holmes et al 1986; Glenn et al 1992). Therefore, we turned to electroporation.

As opposed to other types of bacteria covered in this manual, those discussed in this chapter require an external pH in the

Thomas E. Ward, Bechtel BWXT Idaho, LLC, Idaho National Engineering and Environmental Laboratory, P. O. Box 1625, Idaho Falls, ID, 83415-2203, USA (*phone* +01-208-526-0615;
fax +01-208-526-0828; *e-mail* tew2@inel.gov)

range of 2.5 - 4.5 for growth. However, like most microorganisms, the internal pH of these cells is near neutral (Goulbourne et al 1986). Electroporation creates a disruption of the cell membrane permitting exchange of material, including DNA, between the internal and external environments. Because of this, electroporation of these cells must be performed in neutral solutions, matching their internal pH. However, as mentioned this is outside the range of growth conditions for these bacteria, and they lose viability when held at these pH's. [In our hands, the half lives of these strains varied between 2 hours and 20 hours at pH 7.0 (Glenn et al 1992).] Thus speed is of the essence when using this protocol.

Materials

- Electroporation Apparatus
- 2mm Electroporation Cuvettes (BTX, Biorad, Promega)
- Gel loading pipette tips
- Modified Acidophile Salts (MAS) medium
  - 1mM $(NH_4)_2SO_4$
  - 2mM KCl
  - 0.86 mM K_2HPO_4, pH 3.0 with H_2SO_4
  - 10 mM $MgSO_4$
  - 6.6 mM $CaCl_2$
  - 2.6 mM $FeSO_4$
  - 0.01% (w/v) yeast extract
  - 0.1% (v/v) glycerol

MAS medium is a modification of the medium of Wichlacz and Unz (1981) and is used for cell growth, expression, and selection. Solidified medium is prepared by addition of 1.0% Low EEO agarose (Sigma #A-6013). Tetracycline and chloramphenicol are added to the media at 40 and 70 µg/ml, respectively.

- Wash Buffer, Electroporation Buffer: 1 mM HEPES, pH 7.0

High quality water is essential for electroporation experiments due to the problem of arcing caused by the presence of metal ions in the electroporation solution. Such water can be obtained either from a properly operating still or from a multiple cartridge water purification system. For these experiments, we used a

NANOpure cartridge system from SYBRON/Barnstead, producing 18 meg ohm water.

Procedure

1. Grow cells in MAS medium (above) or other suitable medium, to late log phase, at 37°C if this temperature does not significantly decrease their growth rate, otherwise at 30-32°C.

2. Harvest cells by centrifugation at 5000 rpm, 4°C for 5 minutes, and wash 2-5 times with $1/2$ volume of ice cold 1mM HEPES pH 7.0. The number of washes depends on the metal ion concentration in the growth medium. With MAS medium, two washes are usually sufficient. More washes are needed when using growth media containing higher metal concentrations, in order to prevent arcing during electroporation. From this point on, cells and all solutions should be kept ice cold, and work should be performed as rapidly as possible, consistent with accuracy.

3. Finally, resuspend cells in ice cold 1mM HEPES, pH 7.0, at $10^9 - 10^{11}$ per ml.

4. Aliquot 100 – 200 μl of concentrated cells to pre-chilled electroporation cuvettes. Two-mm cuvettes generally offer a good compromise between ease of use and ability to obtain high enough field strengths. We aliquot the cells using a long, narrow gel-loading pipette tip.

5. Using a gel-loading pipette tip, add appropriate circular plasmid DNA, 0.1-2.0 μg, in sterile, high quality water, keeping the volume of DNA added (1-2 μl) at or below 2% of the volume of cells. Mix gently using the gel-loading pipette tip.

6. Place the cuvette in the electroporation apparatus and discharge a pulse through the mixture of cells and DNA. For these cells, we use settings of 12-15 kV/cm and a 5-10 ms (exponential decay) pulse.

7. Continuing to work as rapidly as possible, dilute the electroporated cells 20-fold into growth medium (at room temperature) and incubate overnight (17-24 hours) at 32°C with shaking to allow expression of the antibiotic resistance pheno-

type. In the case of tetracycline selection, we include tetracycline at a concentration of 1 µg/ml in the expression medium in an attempt to induce expression of the tetracycline resistance genes. We have no experimental evidence concerning whether this is necessary.

8. After expression, plate 200 µl, and appropriate dilutions, of the expression mixture on solid medium containing the selective antibiotic: 40 µg/ml for tetracycline; 70 µg/ml for chloramphenicol, and incubate at 32°C. Transformed colonies should be visible in 3-4 days.

Results

Table 1 shows the results of several experiments. Section A shows that there is significant variability in transformation efficiency among different strains of acidophilic heterotroph. It is suspected that a significant contributor to this variability, as well as to the relatively low transformation efficiency, is the presence of restriction enzymes in these various strains. There have been a number of reports of restriction enzymes isolated from this type of bacteria (Dou et al 1989; Inagaki et al 1990; Inagaki et al 1991; Sagawa et al 1992). The plasmids used in our experiments would not be expected to be protected from restriction enzymes, since they were prepared from *E. coli* cells lacking restriction systems. Section B shows the effects of pre-growth temperature on transformation efficiency. For the two strains tested, growth of the cells at 37°C increases the transformation efficiency approximately tenfold compared to growth at 32°C. Section C shows that the exact structure of the plasmid does not have a major effect on transformation efficiency. However, it is clear that the colE1 replicon present in pBR328 does not work in these cells. In control experiments in which the most readily transformed strain (PW1) was electroporated in the absence of plasmid DNA, or mixed with plasmid DNA but not subjected to an electric pulse, no transformants were obtained.

Table 1. Factors Influencing Electrotransformation Frequency of Acidophilic Heterotrophs (This table is reproduced from Glenn AW, Roberto FF, Ward TE (1992) Transformation of *Acidiphilium* by Electroporation and Conjugation. Can J Microbiol 38:387-393, and used with permission.)

A. Effect of recipient strain on transformation frequency

(plasmid, pRK415; pre-growth, 32 °C; electroporation, 10kV/cm, 5 ms)

	Strain	Transformants/µg DNA	Repeated
	CM3A	$5.3 \pm 2.7 \times 10^{-1}$	3X
	CM5	$8.6 \pm 8.4 \times 10^{1}$	2X
	CM9	$1.4 \pm 0.9 \times 10^{2}$	4X
	CM9A	3.2×10^{0}	1X
	BBW	$1.2 \pm 0.6 \times 10^{1}$	3X
	PW1	$2.1 \pm 0.5 \times 10^{3}$	2X
	PW2	$9.8 \pm 6.3 \times 10^{1}$	7X

B. Effect of pre-growth temperature on transformation frequency

(plasmid, pRK415; electroporation, 10 kV/cm, 5 ms)

Strain	Transformants/ µg DNA		
	21°C	32°C	37°C
CM9A	0	3.2	3.2×10^{1}
PW1	2.7×10^{1}	1.9×10^{3}	2.3×10^{4}

C. Effect of different plasmids on transformation frequency

(strain, PW1; pre-growth, 37 °C; electroporation, 15 kV/cm, 10 ms)

Selection	Plasmid	Transformants/ µg DNA	Repeated
Tetracycline	pL13-4	$2.6 \pm 0.8 \times 10^{4}$	2X
"	pL15-2	$4.6 \pm 2.6 \times 10^{4}$	3X
"	pL17-2	1.5×10^{4}	1X
"	pL19-4	$5.6 \pm 4.2 \times 10^{4}$	2X
"	pRK415	$2.4 \pm 0.5 \times 10^{4}$	9X
"	pLAFR3	$1.4 \pm 1.3 \times 10^{4}$	2X
Chloramphenicol	pL16-1	$2.0 \pm 0.6 \times 10^{4}$	2X
"	pL20-3	$2.0 \pm 1.1 \times 10^{4}$	3X
tet, cam	pBR328	0	3X

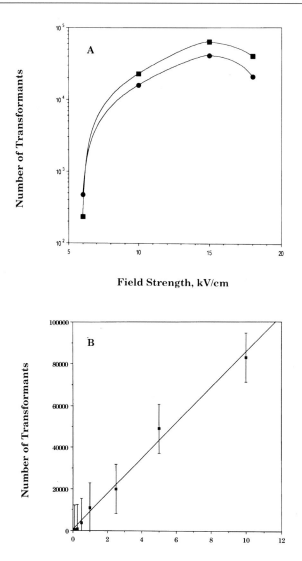

Fig. 1. Electroporation of *Acidocella facilis* strain PW1. Cells were pre-grown at 37°C and plasmid pRK415 was used. Panel A. Effect of field strength and pulse length on transformation frequency. DNA concentration: 2.5 µg/ml. •, 5 ms pulse; ■, 10 ms pulse. Panel B. Effect of DNA concentration on transformation frequency. Electroporation conditions: 10 kV/cm, 5 ms. The bars represent the standard errors of the measurements. The line is a simple linear regression fit to the data, correlation coefficient = 0.99. This figure is reproduced from Glenn AW, Roberto FF, Ward TE (1992) Transformation of *Acidiphilium* by Electroporation and Conjugation. Can J Microbiol 38:387-393, and used with permission.

Panel A of Figure 1, shows that there is a broad maximum of transformation efficiency between 10 and 18 kV/cm field strength and that a 10 ms pulse is slightly more effective than a 5 ms pulse for strain PW1. This does vary between strains, and a 5 ms pulse produces slightly more transformants with some strains (Glenn et al 1992). Little or no cell death was observed using these various treatments. Panel B shows that the number of transformants increased linearly with increasing DNA concentration up to 10 μg/ml. The original plasmid could be recovered from the transformants (Glenn et al 1992).

Experiments with cells harvested in early log, mid log, late log/early stationary and late stationary phases of growth did not result in significant differences in electroporation frequency (data not shown). Likewise, use of various low-ionic-strength buffers at pH 7.0, containing either HEPES or potassium phosphate, with and without sucrose and $MgCl_2$ did not have a significant effect on transformation efficiency. On the other hand, electroporation in 5 mM potassium phosphate, pH 3.0 resulted in a 5000 fold decrease in transformation efficiency and a 5 fold decrease in cell survival compared to the same experiment performed at pH 7.0 (Glenn et al 1992).

Comments

- The only named species used in our studies (Glenn et al 1992) was *Acidiphilium facilis* (ATCC 35904, Wichlacz strain PW2). We also used a number of independent isolates from acid mine drainage environments, which are physiologically similar to this species. Since that time, new results (Kishimoto et al 1995) have led to the division of the previous genus *Acidiphilium* into two genera, and the renaming of the genus containing what previously was *Acidiphilium facilis* as *Acidocella*. We believe the conditions described here can be used to electrotransform most, if not all, of these acidophilic, heterotrophic, Gram-negative eubacteria. This procedure has been successfully used by others (Ghosh et al 1997).

- A variety of growth media are used by different researchers working with acidophilic bacteria, usually in an attempt to mimic the ionic conditions of the environments from which the bacteria were collected. The only complication caused by these different growth media is the requirement to wash the cells thoroughly before electroporation to remove as many of the metal ions as possible. Otherwise arcing during the pulse can readily occur.

- Pre-growth of the cells at a temperature above that which is optimal for growth significantly increases the frequency of electrotransformation. We postulate that this is due to alteration of the lipid composition of the cell membrane caused by growth at the elevated temperature, as has been observed for other bacteria (Harwood and Russell 1984; Neidleman 1987), which results in enhanced ability to take up DNA. This strategy may not be feasible for all acidophilic heterotrophs, some of which grow poorly at 37°C. For these strains, it is expected that growth at temperatures such as 34-35°C would result in increased transformation efficiency, although probably not to the extent observed after growth of tolerant strains at 37°C. A thorough study of the effects of a variety of elevated growth temperatures on electrotransformation frequency has not been performed.

- As mentioned, the stage of growth in which the cells are harvested has only a minor effect on transformation efficiency. Cells are harvested in late log phase to make optimum use of medium.

- Incubation for expression at 32°C for up to 20 hours resulted in no more than a 2-3 fold increase in cell number.

- We have not had success using ampicillin and an ampicillin resistance gene as a selective system with these organisms, although the ampicillin did appear to be stable in acidic medium, i.e. the growth of cells was inhibited. However, no resistant colonies appeared on amp plates following electroporation with a plasmid carrying both amp and tet resistance genes, whereas colonies did form when an aliquot from the same expression mixture was spread on tet plates. We attribute this to the fact that ampicillin resistance is mediated by a beta-lactamase which is excreted into the medium. The gene coding for this enzyme is from an enteric bacterium, and thus the activity of this enzyme would be expected to be optimal near neutral pH. Since the external medium for growth of these cells is pH 3.0-3.5, any beta-lactamase excreted would probably not be active.

- Two different tetracycline resistance genes (from pBR328 and pRK415) with their endogenous (enteric) promoters, and a chloramphenicol resistance gene from pBR328, have been successfully used as selectable markers. The activity of the enzymes responsible for resistance which are produced in these acidophilic cells has not been measured. Our functional test requires that enough activity be produced to provide resistance to the specified levels of antibiotics, which completely prevent growth of non-transformed cells.

- A broad-host-range origin of replication or an origin from an endogenous acidophile plasmid is required for a plasmid to replicate in these cells.

- Two electroporation instruments from BTX Corp., San Diego, Calif. were used for these experiments, a Transfector 100 and an ECM 600. Other commercially available instruments should work as well. No experiments have been performed using square wave pulses.

- Because of the aforementioned pH incompatibility, we have not tried to store frozen electrocompetent cells for later use, as is commonly done with *E. coli*.

Acknowledgements. The excellent technical assistance of Debbie Bulmer, Anne Glenn and Anna (Wilhite) Lugar is greatfully acknowledged. I also thank Debbie Bulmer for a careful reading of the manuscript. This work was supported through the INEL Long Term Research Initiative Program under U. S. Department of Energy Idaho Operations Office Contract No. DE-AC07-76IDO1570 to EG&G Idaho, Inc.

References

Dou D, Inagaki K, Kita K, Ohshima A, Hiraoka N, Kishimoto N, Sugio T, Tano T (1989) Restriction endonuclease AfaI from *Acidiphilium facilis*, a new isoschizomer of RsaI: purification and properties. Biochim Biophys Acta 1009(1):83-86

Ghosh S, Mahapatra NR, Banerjee PC (1997) Metal resistance in *Acidocella* strains and plasmid-mediated transfer of this characteristic to *Acidiphilium multivorum* and *Escherichia coli*. Appl Environ Microbiol 63(11):4523-4527

Glenn AW, Roberto FF, Ward TE (1992) Transformation of *Acidiphilium* by Electroporation and Conjugation. Can J Microbiol 38:387-393

Goulbourne E, Jr., Matin M, Zychlinsky E, Matin A (1986) Mechanism of "delta" pH maintenance in active and inactive cells of an obligately acidophilic bacterium. J Bacteriol 166:59-65

Harrison AP, Jr. (1984) The acidophilic thiobacilli and other acidophilic bacteria that share their habitat. Ann Rev Micro 38:256-292

Harrison AP, Jr. (1989) Genus *Acidiphilium*. In: J. T. Staley (ed.). Bergey's Manual of Systematic Bacteriology, Vol. 3. Williams and Wilkins, Baltimore, pp 1863-1868

Harwood JL, Russell NJ (1984). Lipids in Plants and Microbes. Allen & Unwin, London

Holmes DS, Yates JR, Lobos JH, Doyle MV (1986) Strategy for the establishment of a genetic system for studying the novel acidophilic heterotroph *Acidiphilium organovorum*. Biotech Appl Biochem 8:1-11

Inagaki K, Kobayashi F, Dou DX, Nomura Y, Kotani H, Kishimoto N, Sugio T, Tano T (1990) Isolation and identification of restriction endonuclease Asp35HI from *Acidiphilium* species 35H. Nucleic Acids Res 18(20):6155

Inagaki K, Ito T, Sagawa H, Kotani H, Kishimoto N, Sugio T, Tano T, Tanaka H (1991) AcpI, a novel isoschizomer of AsuII from *Acidiphilium cryptum* 25H, recognizes the sequence 5'TT/CGAA3'. Nucleic Acids Res 19(22):6335

Kishimoto N, Kosako Y, Wakao N, Tano T, Hiraishi A (1995) Transfer of *Acidiphilium facilis* and *Acidiphilium aminolytica* to the Genus *Acidocella* gen. nov., and Emendation of the Genus *Acidiphilium*. System Appl Microbiol 18:85-91

Neidleman SL (1987) Effects of Temperature on lipid unsaturation. Biotechnology and Genetic Engineering Review 5:245-268

Sagawa H, Takagi M, Nomura Y, Inagaki K, Tano T, Kishimato N, Kotani H, Nakajima K (1992) Isolation and identification of restriction endonuclease Aor51HI from *Acidiphilium organovorum* 51H. Nucleic Acids Res 20(2):365

Wichlacz PL, Unz RF (1981) Acidophilic, heterotrophic bacteria of acidic mine waters. Appl Environ Microbiol 41:1254-1261

Abbreviations

amp	ampicillin
cam	chloramphenicol
tet	tetracycline

Acetobacter xylinum – Biotechnology and Food Technology

ROBERT E. CANNON

▨ Introduction

Acetobacter xylinum is a Gram negative soil bacterium that synthesizes and secretes cellulose as part of its metabolism of glucose. Cellulose is the most abundant natural polymer on Earth, and is the major constituent of the cell wall of plants. The cellulose produced by *A. xylinum* is structurally and chemically identical to cellulose found in higher plants, and in addition, is not contaminated by lignins or other cellulosic derivatives. It is for this reason that *Acetobacter xylinum* serves as a potential model organism for the study of cellulose biosynthesis.

Cellulose from *Acetobacter* has potential commercial applications in a variety of industries. The fibrils of cellulose that are synthesized are much narrower than wood fibers, on the order of 0.1 μm, which makes bacterial cellulose a possible material for high quality paper, an emulsifier for paints, and a food additive. Bacterial cellulose has also been used as a skin substitute to protect burned skin, and for ultra filtration technology.

Acetobacter xylinum synthesizes cellulose from glucose through a number of enzymatically catalyzed reactions. Glucose is converted into glucose-6-phosphate by Glucokinase. Glucose-6-phosphate is then converted into glucose-1-phosphate by Phosphoglucomutase. Glucose-1-phosphate is then converted into UDP-glucose by UDP-glucose pyrophosphorylase, and finally UDP-glucose is converted to cellulose by Cellulose synthase. Cellulose is considered a secondary metabolite, which

Robert E. Cannon, University of North Carolina at Greensboro, Department of Biology, P. O. Box 26174, Greensboro, North Carolina, 27402-6174, USA (*phone* +01-336-256-0071; *fax* +01-336-334-5839; *e-mail* recannon@uncg.edu)

the bacterium synthesizes after major metabolic needs are met. Genetic and molecular studies have shown that cellulose synthesis is under operon control, and that the process is regulated by a cyclic diguanylic activator of the cellulose synthase that functions at the plasma membrane of the bacterial cell to allow proper polymerization of the glucose molecules into the growing and extruded cellulose microfibrils.

Since *A. xylinum* is a potential model organism to study cellulose synthesis, it would be useful to have methodologies to study it genetically. The three methods of genetic exchange that have been observed in bacteria are transduction, conjugation, and transformation. At present, there are no known bacteriophages that specifically infect this bacterium so phage-mediated transduction is not an option. It is possible to transfer genetic material into *A. xylinum* by conjugation with broad host range plasmids from other bacteria such as *Escherichia coli*, but little progress has been made in developing a conjugation system in *A. xylinum*. Thus, we pursued the possibility that transformation might be the most appropriate method for inserting genetic information into *Acetobacter* as a prelude to more extensive genetic analyses. When standard chemical methods of transformation failed, electroporation was found to be successful.

Outline

1. Mid-log *A. xylinum* cells harvested and resuspended in 15% glycerol at cell density of 10^{10} cells/ml.

2. Cell suspensions divided into 200 ml aliquots and frozen at -70°C.

3. 2 μl of plasmid DNA added to cells.

4. Cell/DNA mixture placed in chilled 0.2 cm electroporation cuvette.

5. Cells pulsed for 6-8 ms at field strength of 12.5 kV/cm.

6. Cells transferred to 800 μl of *Acetobacter* culture medium and shaken overnight at room temperature.

7. Cells plated on solid *Acetobacter* culture medium to select for transformants.

Materials

Equipment   Gene Pulser (Bio-Rad Laboratories, Richmond, CA)

Media and   – Acetobacter xylinum culture medium
chemicals

Glucose	20 g/l
Yeast Extract	5 g/l
Peptone	5 g/l
Dibasic Sodium Phosphate	2.7 g/l
Citric Acid	1.15 g/l

Bacterial   – Cellulose-negative *Acetobacter xylinum* strain derived from
strains and   *A. xylinum* ATCC 23769
plasmids   – *Escherichia coli* strains HB101 and MM294
             – Plasmids - pUCD2 and pRK248

Procedure

1. Collect mid-log phase *A. xylinum* cells by centrifugation, wash twice with cold 1 mM HEPES buffer (pH 7.0), and re-suspend in 15% glycerol to a final cell density of 10^{10} cells/ml. Distribute cell suspensions into 200 µl aliquots in microfuge tubes and freeze at -70 °C.

2. Add two microliters of plasmid DNA varying from 20 pg to 0.2 µg (100 pg/ml to 1 µg/ml) to 200 µl of previously concentrated cells in a cold microfuge tube. Place the cell/DNA mixture in a chilled 0.2 cm electroporation cuvette, which is then placed between the electrodes for application of the pulse.

3. After pulsing at 25 µF capacitance, 2.5 kV, and 400 ohms resistance for a generated pulse length of 6-8 ms and field strength of 12.5 kV/cm, transfer cells to 800 µl of *A. xylinum* culture medium.

4. Shake cells gently overnight at room temperature to permit expression of plasmid genes.

5. Dilute and spread cells on solid *Acetobacter* medium containing tetracycline (15 µg/ml) to select for transformants and on nonselective medium (no antibiotic present) to assay survival of cells from the electroporation process.

6. Plasmid transformation via electroporation may be confirmed by restriction enzyme analysis of plasmid DNA isolated from transformants and compared with CsCl-purified plasmid and by transformation of an *E. coli* host (MM294) with plasmid isolated from *A. xylinum* primary transformants. Cells pulsed in the absence of DNA yielded no transformants when plated on selective media.

Results

Acetobacter xylinum may be efficiently transformed using high-voltage electroporation. The transformation efficiency (transformants/ µg DNA) remains fairly constant depending on the particular plasmid used and ranges from 10^5 to 10^7 transformants/ µg DNA. Frequency of transformation (transformants/ survivor) increases linearly with increasing DNA concentration. Some transformation without pulsing occurs at high DNA concentration (1.0 µg/ml) indicating that some natural competence may occur in *Acetobacter xylinum*, although the factors influencing the development of such competence have not been identified. Restriction analysis of plasmid DNA isolated from *A. xylinum* transformants suggests that plasmids do not undergo deletion or rearrangement as a consequence of the electroporation process.

References

Hall P et al. (1992) Transformation of *Acetobacter xylinum* with plasmid DNA by electroporation. Plasmid 28: 194-200

Cannon R, Anderson S (1991) Biogenesis of bacterial cellulose. CRC Critical Reviews in Microbiology 17: 435-447

Toda K et al. (1997) Cellulose production by acetic-acid resistant *Acetobacter xylinum*. J. Fermentation and Bioengineering 84: 228-231

Electrotransformation of *Sphingomonas paucimobilis*

Isabel Sá-Correia and Arsénio M. Fialho

Introduction

Bacterial strains of the new genus *Sphingomonas* (Yabuuchi et al 1990) are relatively ubiquitous in soil, water and sediments and have broad catabolic capabilities (Fredrickson et al. 1995; Dutta et al. 1998). Beside the high potential of *Sphingomonas* spp. for bioremediation and waste treatment, different strains of this genus produce at least eight extracellular acid heteropolysaccharides that have similar but not identical structures (Chandrasekaran and Radha 1995). These polysaccharides, the sphingans, exhibit properties which make them candidates for food and industrial applications such as thermoreversible gel formation and solution viscosity (Chandrasekaran and Radha 1995). The producing bacteria were originally classified into diverse genera (*Pseudomonas, Alcaligenes, Azotobacter* and *Xanthobacter*) but were later shown to be closely related to each other and to members of the genus *Sphingomonas*; most of them were identified as *Sphingomonas paucimobilis* (Pollock 1993). This is the case for the industrial strain ATCC 31461 (formerly *Pseudomonas elodea*) (Kang and Veeder 1982) that synthesize, in high yields, gellan gum. Gellan is currently produced by large scale aerobic fermentation for use as a gelling agent in foods and microbiological and plant tissue culture media and in other non-food applications such as personal care products and controlled delivery of drugs.

✉ Isabel Sá-Correia, Centro de Engenharia Biológica e Química, Instituto Superior Técnico, Av. Rovisco Pais, Lisboa Codex, 1049-001, Portugal (*phone* + 351-2-8417682; *fax* + 351-2-8480072; *e-mail* pcisc@alfa.ist.utl.pt) Arsénio M. Fialho, Centro de Engenharia Biológica e Química, Instituto Superior Técnico, Av. Rovisco Pais, Lisboa Codex, 1049-001, Portugal; *e-mail* pcfialho@alfa.ist.utl.pt)

The commercial utility of gellan has been a stimulus for studying its biosynthesis. The cloning and the functional analysis and expression profiles of genes essential for gellan synthesis are indispensable for the genetic and environmental manipulation of its biosynthetic pathway. This will allow the development of new polysaccharides, with distinct structure and consequently diverse physical properties and commercial applications (Fialho et al. 1999). One of the critical steps in the cloning of gellan genes was the development of methods to introduce recombinant plasmids into *S. paucimobilis* ATCC 31461. The transformation of this strain with plasmid DNA (above 10 kb) by the $CaCl_2$ or RbCl - heat shock technique was unsuccessful but plasmid mobilization from *Escherichia coli* by triparental mating was successfully used (Fialho et al 1991). In order to open up the possibility of cloning gellan genes directly into *S. paucimobilis* for complementation experiments without the need to use *E. coli* as an intermediate host, as in the case of triparental conjugative plasmid transfer, an electrotransformation protocol was developed for this organism. This protocol is rapid, reliable, leads to optimal transformation efficiency and can be used in the cloning of the gellan genes (Monteiro et al 1992). Protocol optimization concerned the preparation of cells for electrotransformation (selection of growth medium and growth phase), the selection of optimal cell and plasmid DNA concentrations in the electrotransformation mixture, and the field strength and pulse length (Monteiro et al 1992). The effect of plasmid size on electrotransformation frequency was also analysed (Monteiro et al 1992).

Materials

- Electroporation apparatus (Gene Pulser with a Pulse Controller, Bio-Rad Laboratories, Hercules, CA, USA)   **Equipment**
- 0.2 cm gap electroporation cuvettes

Strains and plasmids

Bacterial strains and plasmids

Strain or plasmid	Relevant characteristics
Sphingomonas paucimobilis	
ATCC 31461	Gel$^+$, Rif^s
R40	Gel$^+$, Rif^r
RP10	Gel$^-$, Rif^r
Escherichia coli DH1	*rec^A*, *hsdR*
Cloning vectors	
pKT240	(12.9 kb), RSF1010, Km^r, Ap^r
pJRD215	(10.2 kb), RSF1010, Km^r, Sm^r, *cos*
Recombinant plasmids	
- Plasmids from a *S. paucimobilis* R40 genomic library constructed in pKT240, with inserted DNA varying in the range 2 - 12 kb, and hosted in *E. coli* DH1	(range 15 - 25 kb), Km^s, Ap^r
- Plasmids from a *S. paucimobilis* R40 cosmid genomic library in pJRD215 with inserted DNA varying in the range 10 - 25 kb and hosted in *E. coli* DH1	(range 20 - 35 kb), Km^r, Sm^r, *cos*

Growth media

LB medium (g/l in destilled water)

Peptone (Difco, Detroit, MI, USA)	10
Yeast extract (Difco)	5
NaCl	5
pH 7.0, sterilize in the autoclave	

S medium (g/l in distilled water)

Na_2HPO_4	10
KH_2PO_4	3
K_2SO_4	1
NaCl	1
Yeast extract (Difco)	1
Casamino acids (Difco)	1
$MgSO_4.7H_2O$	0.2
$CaCl_2.2H_2O$	0.01
$FeSO_4.7H_2O$	0.001
glucose	20
pH 7.5	

$MgSO_4$, $CaCl_2$ and $FeSO_4$ stock solutions (10x) are prepared and sterilized, separately, by filtration. The other ingredients are poured into three bottles [(salts), (yeast extract, casamino acids) and (glucose)] and sterilized, separately, by autoclaving.

LB-and S-solid media are prepared by adding 2% (w/v) agar (Iberoagar, Portugal).

Antibiotics are filter sterilized and added to selective media, depending on the marker(s) of plasmids to be introduced by electrotransformation. Final concentrations in solid (liquid) media (µg/ml) are:

ampicillin	150 (75)
kanamycin	100 (50)
tetracycline	25 (12.5)

Buffers and solutions

- Wash buffer
  - 1 mM Tris base in water
  - pH 8.0; adjust with 1 M HCl
- Electrotransformation buffer
  - 10% (v/v) glycerol (Merck, Darmstadt, Germany) in water
  - pH 4.5; adjust with 1 M NaOH
- Saline solution: 0.9% (w/v) NaCl, sterilize in the autoclave

▨ Procedure

Preparation of *S. paucimobilis* cells for electrotransformation

1. Grow the gellan producing or non-producing strains of *S. paucimobilis* in 1 l LB medium in 2 l Erlenmeyer flasks at 30°C with orbital agitation (150 rev min^{-1}) until the late exponential phase is reached (initial OD_{640nm} 0.1 ± 0.02 by inoculation with a liquid pre-inoculum and cells are grown up to OD_{640nm} 1.3 ± 0.05, corresponding to approximately 10 hours of incubation).

2. Recover the cells by centrifugation at 5400xg for 30 minutes at 4°C.

3. Wash the cell pellets twice with 250 ml of the wash buffer and once with 250 ml of the electrotransformation buffer.

4. Gently resuspend the cell pellets in 4 ml of the electrotransformation buffer (corresponding to approximately 1-2 x 10^{10} CFU/ml). Asseptically transfer 500 µl aliquots of each cell suspension to sterile cryotubes and rapidly freeze and store the cell suspensions at -70°C until needed or use them immediately for electrotransformation. No significant loss of cell viability occurs for at least up to 9 months of storage at -70°C.

Plasmid isolation and quantification

1. Extract plasmids (cloning vectors or recombinant plasmids from genomic libraries) from *E. coli* host strains (e.g. *E. coli* DH1) by lysis with sodium dodecyl sulphate followed by phenol/chloroform extractions (Sambrook et al 1989). Plasmid concentration can be estimated from a densitometry scan of the negatives of electrophoresis gel photographs (Monteiro et al 1992). In order to increase electrotransformation frequency it is advisable to carry out the microdialysis on membranes of the mini-prep plasmid DNA solution or caesium chloride purification of plasmid DNA (Sambrook et al 1989).

Electrotransformation of *S. paucimobilis*

1. Remove an aliquot of the frozen cell suspensions and leave to stand on ice until thawed. This step is unnecessary if freshly prepared cells are used.

2. Gently transfer 100 μl of the chilled cell suspension to a pre-chilled 0.2 cm gap electroporation cuvette (avoiding the formation of air bubbles). The use of ultrasonic vibrations to destroy cell aggregates before electrotransformation may improve process efficiency. In general, the incubation of cell suspensions in a sonicator bath (e. g. Elma, Transsonic 460, Singen, Germany) for 5 minutes may lead to increased number of electrotransformants.

3. Add 10 μl of a mini-prep plasmid DNA solution (diluted in water) to obtain a final concentration of approximately (or above) 500 ng/ml for plasmids of around 13 kb (pKT240). For bigger (or smaller) plasmids, DNA concentration should be calculated in order to maintain at least two plasmids per CFU in the electrotransformation mixture. For lower concentrations, the number of electrotransformants/ml of electroporation mixture is an exponential function of plasmid concentration.
Prepare controls in which plasmid DNA is omitted in order to assess the percentage of spontaneous antibiotic resistant cells.

4. Expose the electroporation cell suspension to a field strength of 12.5 kV/cm, delivered with a 25 μF capacitor, and a pulse controller set at 400 Ω (pulse length range between 7.5 - 9.0 msec). Electrotransformation can be carried out without pre-incubation of cells with plasmid DNA because incubation for up to 20 minutes has no effect on the frequency of electro-transformation.

5. Immediately dilute the cell suspension with 1 ml of LB medium and incubate in a sterile tube with orbital agitation (250 rev min^{-1}) at 30°C for 2 hours.

6. After serial dilution in sterile saline solution, spread the cells on S-agarized medium supplemented with ampicillin (for pKT240 or recombinant plasmids constructed in pKT240)

or kanamycin (for pJRD215 or recombinant cosmids of the cosmid genomic library) and incubate the plates at 30°C for 96 hours in order to obtain electrotransformant colonies and to compare their mucoid phenotype.

Results

Using optimal electrotransformation conditions [1-2 x 10^{10} CFU/ ml of cells grown in a medium leading to low gellan production and harvested in the late exponential phase; a concentration of plasmid equivalent to at least two plasmids/CFU, a pulse strength of 12.5 kV/cm and a pulse length in the range 7 - 9 msec (using the 400 Ω parallel resistor)] it was possible to obtain a number of electrotransformants of non-mucoid *S. paucimobilis* ranging from 3 x 10^7/ml to 8 x 10^2/ml of the electrotransformation mixture, when cosmid pJRD215 (10.2 kb) or a recombinant plasmid of 25 kb were used, respectively (Monteiro et al 1992). This corresponds to electrotransformation frequencies of approximately 3 x 10^{-3} or 8 x 10^{-8} and electrotransformation efficiencies of 8 x 10^7 and 8 x 10^2/µg DNA for 10.2 kb or 25 kb plasmids, respectively. The exponential decrease of electrotransformation frequency with the increase of plasmid size (Figure 1) was virtually independent of the broad-host-range cloning vector used (pKT240 or pJRD215) (Monteiro et al 1992). Experimental evidences (Monteiro et al 1992) indicated that the strong influence of plasmid size on the electrotransformation frequency is basically due to the different number of targets for restriction enzymes that can be present in plasmid DNA. Among them, it was found that when plasmid DNA bigger than 13 kb was extracted from *E. coli* host strains and introduced into *S. paucimobilis* strains, with an active modification/restriction system, the electrotransformation frequency was drastically reduced compared with that possible using plasmids extracted from *S. paucimobilis*, where they were introduced by conjugation (Monteiro et al 1992).

No dramatic differences were observed in the electrotransformation frequency of gellan-producing and non-producing strains when cells for electrotransformation were grown in media leading to low gellan production. Nevertheless, under identical conditions, and despite plasmid size, the number of electrotrans-

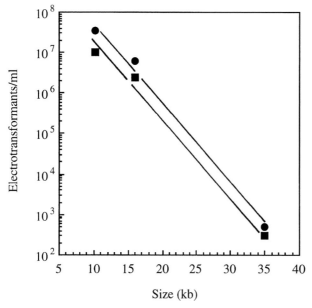

Fig. 1. Effects of the plasmid size on the number of electrotransformants per ml of the electrotransformation mixture of gellan-gum producing *Sphingomonas paucimobilis* (formerly *Pseudomonas elodea*) R40 (■) and of the non-producing spontaneous variant strain RP10 (●) with cosmid pJRD215 (10.2 kb), or recombinant cosmids of 16 kb and 35 kb. Electrotransformation conditions: 2.5 kV/cm, $1 - 2 \times 10^{10}$ CFU/ml, 660, 1050 and 2300 ng/ml of plasmid DNA, respectively, in order to maintain the relationship of approximately three to six plasmids per CFU in the electrotransformation mixture. (reproduced with permission from Monteiro et al 1992)

formants that can be obtained with mucoid cells is at least one half the number of non-mucoid cells that can be electrotransformed (Figure 1). This difference increases when cells for electrotransformation are grown in media that allow the production of large amounts of gellan. Both mucoid and non-mucoid cells of *S. paucimobilis* prepared for electrotransformation form big aggregates that limit the number of transformant colonies that can be obtained. Under optimal conditions, sonication may slightly increase the number of electrotransformants by reducing the number or size of cell aggregates without a drastic decrease of cell viability.

Comments

Even under optimized conditions, the electrotransformation of gellan-gum producing or non-producing strains derived from *S. paucimobilis* ATCC 31461 is strongly limited by the size of plasmid DNA. Above 30 - 35 kb, no transformants (or a very small number) can be obtained even when the electrotransformation cell suspension is directly plated, without dilution, on selective S-agar plates. This means that this technique fails if the objective is the identification of chromosomal DNA regions necessary for gellan biosynthesis by the introduction into Gel⁻ mutants of a cosmid genomic library packed in λ particles and introduced into *E. coli* by transduction (expected sizes in the range 38.5-52 kb). Nevertheless, it is possible to electrotransform Gel⁻ mutants with recombinant plasmids with DNA inserts up to 15 kb when the cloning vectors usually used in the genetic engineering of Gram-negative bacteria (with approximately 10 kb) are used. However, the heterogeneity of recombinant plasmid sizes of a representative genomic library should be strictly restricted in order to avoid the overrepresentation of electrotransformants harbouring the smaller plasmids of the gene bank.

As observed with the electrotransformation frequency, the mobilization frequency of recombinant plasmids into *S. paucimobilis* decreases, exponentially, with plasmid size and the frequency of conjugal transfer of plasmids is higher into non-mucoid than into mucoid spontaneous variants or mutants (Fialho et al 1991). However, under optimal conditions, it is possible to obtain a reasonable number of transconjugants with slightly bigger plasmids that can give rise to electrotransformants (Fialho et al 1991, Monteiro et al 1992).

Electrotransformation proved to be a valuable alternative to conjugal mating in the search for complementation of gellan mutations without the need to use *E. coli* as an intermediate host. Moreover, this technique avoids the use of rifampicin necessary to select *S. paucimobilis* transconjugants (Fialho et al 1991) as the expense, associated with the addition of this antibiotic is eliminated. In addition, since the frequency of the mucoid phenotype spontaneous variation is stimulated (Martins et al 1996) and gellan synthesis is reduced (Fialho et al 1991) in rifampicin supplemented plates, electrotransformation renders easier the task of searching for Gel⁻ mutant complementation.

The electrotransformation procedure described here can also be applied to introduce plasmid DNA in other strains of *S. paucimobilis* relevant in Biotechnology and Food Technology, as well as in other exopolysaccharide producing Gram negative bacterial species, after the necessary adaptations. This electrotransformation protocol was used by Vartak et al (1995) to introduce, into a poly-β-hydroxybutyrate deficient mutant of *S. paucimobilis* ATCC 31461, a recombinant plasmid constructed in pBluescript II SK (+) (3 kb) to construct a glucose-6-phosphate dehydrogenase insertion mutant. The plasmid size (10.6 kb), used in this work, allowed the authors to grow the cells for electroporation overnight on agar plates and to use them successfully, for the preparation of electrocompetent cells (Vartak et al 1995).

References

Chandrasekaran R, Radha A (1995) Molecular architectures and functional properties of gellan gum and related polysaccharides. Trends Food Sci Technol 6:143-148

Dutta TK, Selifonov SA, Gonsalus IC (1998) Oxidation of methyl-substituted naphtalenes: pathways in a versatile *Sphingomonas paucimobilis* strain. Appl Environ Microbiol 64:1884-1889

Fialho AM, Monteiro GA, Sá-Correia I (1991) Conjugal transfer of recombinant plasmids into gellan gum producing and non-producing variants of *Pseudomonas elodea* ATCC 31461. Lett Appl Microbiol 12:85-87

Fialho AM, Martins LO, Donval ML, Leitão JH, Ridout MJ, Jay AJ, Sá-Correia I (1999) Structures and properties of gellan polymers produced by *Sphingomonas paucimobilis* ATCC 31461 from lactose compared with those produced from glucose and from cheese whey. Appl Environ Microbiol 65:2485-2491

Fredrickson JK, Balkwill DL, Drake GR, Romine MF, Ringelberg DB, White DC (1995) Aromatic-degrading *Sphingomonas* isolates from the deep subsurface. Appl Environ Microbiol 61:1917-1922

Kang KS, Veeder GT (1982) Gellan polysaccharide S-60 and bacterial fermentation process for its preparation. US patent 4,326,053

Martins LO, Fialho AM, Rodrigues PL, Sá-Correia I (1996) Gellan gum production and activity of biosynthetic enzymes in *Sphingomonas paucimobilis* mucoid and nonmucoid variants. Biotechnol Appl Biochem 28:47-54

Monteiro GA, Fialho AM, Ripley SJ, Sá-Correia I (1992) Electrotransformation of gellan gum-producing and non-producing *Pseudomonas elodea* strains. J Appl Bacteriol 72:423-428

Pollock TJ (1993) Gellan-related polysaccharides and the genus *Sphingomonas*. J Gen Microbiol 139:1939-1945

Sambrook J, Fritsch EF, Maniatis T (1989) Molecular cloning. A Laboratory Manual. Cold Spring Harbor Laboratory Press, New York

Vartak NB, Lin CC, Cleary JM, Fagan MJ, Saier MH (1995) Glucose metabolism in *Sphingomonas elodea*: pathway engineering via construction of a glucose-6-phosphate dehydrogenase insertion mutant. Microbiology 141:2339-2350

Yabuuchi E, Yano I, Oyaizu H, Hashimoto Y, Ezaki T, Yamamoto H (1990) Proposals of *Sphingomonas paucimobilis* gen. nov. and comb. nov., *Sphingomonas parapaucimobilis* sp. nov., *Sphingomonas yanoikuyae* sp. nov., *Sphingomonas adhaesiva* sp. nov., *Sphingomonas capsulata* comb. nov., and two genospecies of the genus *Sphingomonas*. Microbiol Immunol 34:99-119

Abbreviations

CFU	colony forming units
Gel$^+$/Gel$^-$	gellan producer/non-producer
Ap^r	ampicillin resistance
Km^r/Km^s	kanamycin resistance/sensitivity
Rif^r/Rif^s	rifampicin resistance/sensitivity
Sm^r	streptomycin resistance
msec	milli-second

Glossary

frequency and the efficiency of electrotransformation
The frequency and the efficiency of electrotransformation are the ratio of electrotransformants per survivors (CFU) after electrotransformation and the number of electrotransformants per µg of plasmid DNA, respectively

Bacillus amyloliquefaciens - Production Host for Industrial Enzymes

JARI VEHMAANPERÄ

Introduction

Bacillus amyloliquefaciens is a widely used host for production of industrial enzymes (Aunstrup et al. 1979). However, the species does not become physiologically competent for DNA transformation, unlike the related species *B. subtilis*. Therefore, alternative methods for introduction of plasmids into *B. amyloliquefaciens* have been developed. These include the protoplast transformation (Vehmaanperä 1988) and the electroporation methods (Vehmaanperä 1989). The latter is described below.

The electroporation method enables transformation of both laboratory and industrial *B. amyloliquefaciens* strains with recovery of up to 10^5 transformants/ μg DNA. The protocol also works for *B. subtilis*, yielding upto 10^4 transformants/ μg DNA.

Materials

The method has been developed using the Gene Pulser™ electroporation apparatus and the accessory cuvettes (BioRad, Richmond, CA, USA). Electroporation devices supplied by other manufacturers may require some optimization of the pulse conditions.

– LBSP   **Media and**
  – 10 g/l Bacto tryptone   **buffers**
  – 5 g/l Bacto yeast extract

Jari Vehmaanperä, VTT Biotechnology, Tietotie 2, Espoo, 02044, Finland (*phone* +358-9-4561; *fax* +358-9-455 2103; *e-mail* jari.vehmaanpera@vtt.fi.)

- 5 g/l NaCl
- 250 mM sucrose
- 50 mM K-phosphate (pH 7.2)
- LBSPG
  - LBSP
  - 10% (v/v) glycerol
- SHMG
  - 250 mM sucrose
  - 1 mM Hepes (pH 7.0)
  - 1 mM $MgCl_2$
  - 10 % (v/v) glycerol
- LB agar
  - 10 g/l Bacto tryptone
  - 5 g/l Bacto yeast extract
  - 5 g/l NaCl
  - 18 g/l Bacto agar (pH 7.0)

Procedure

Preparation of cells

1. Inoculate 50 ml of LBSP in a 500 ml Erlenmeyer flask with cells from a single colony grown on a LB agar plate.
   It seems to be important to pre-adapt the *B. amyloliquefaciens* cells in the sucrose containing LBSP medium to the osmotic pressure caused by the sucrose in the electrotransformation buffer SHMG (see below).

2. Grow with vigorous aeration to late exponential phase at 37 °C (OD_{600} ~1.0 / $Klett_{60}$~150).

3. Reinoculate 4 x 200 ml portions (in 1 litre flasks) of pre-warmed LBSP each with 10 ml of the grown culture.

4. Grow with vigorous aeration to OD_{600} ~1.0/ $Klett_{60}$~150 (2-5 x 10^8 cells/ml; about 2 - 3 h) at 37 °C.

5. Chill the cultures in an ice-water bath for 10 min.

6. Harvest the cells by centrifugation at 10 000 g for 5 min at 4 °C using precooled rotor and tubes. Discard the supernatant.

7. Wash the cells three times with about 200 ml of ice-cold SHMG. Pellet the cells between the washes by centrifuging at 10 000 g for 5 min at 4 °C.

8. Resuspend the cells in 4 - 8 ml of cold SHMG and dispense 0.5 ml aliquots in microtubes.

9. Freeze the cells by placing the tubes in a - 70 °C freezer. Do not freeze the cells in dry ice-ethanol bath or in liquid nitrogen. The cells remain transformable for at least several months.

Electrotransformation

1. Thaw 0.5 ml of frozen cells quickly in a 37 °C water bath.

2. Add upto 10 µl of DNA in TE or in SHMG, containing 10 ng - 1 µg of DNA, mix shortly by vortexing.

3. Transfer the cell suspension with DNA to a 0.2 cm electroporation cuvette, and keep in an ice-water bath for 20 min.

4. Apply a single pulse (1500 V = 7.5 kV/cm for *B. amyloliquefaciens*, 2500 V = 12.5 kV/cm for *B. subtilis*; 25 µF for both species)

5. Immediately dilute the cell suspension 1:10 in LBSPG at room temperature.

6. Incubate with vigorous aeration at 37 °C for 60 min for phenotypic expression of the antibiotic resistance marker(s) on the plasmid.

7. Plate on LB-agar supplemented with the antibiotic for selection of the transformants. Clones transformed with pUB110 are preferably selected on 1 - 5 µg phleomycin/ml (Cayla, Toulouse, France) rather than on kanamycin (10 µg/ml), to reduce the background growth.

8. Incubate at 37 °C for 1 - 2 d.

The method has resulted in transformation frequencies of about 10^5 transformants/µg plasmid DNA for *B. amyloliquefaciens*, and about 10^4/µg for *B. subtilis*. With ligation mixtures, the efficiency is about 10-100 -fold reduced.

Troubleshooting

If the transformation frequency in *B. amyloliquefaciens* is much lower than expected, the reason may be an unmethylated *Bam*HI site on the plasmid, which is cleaved by the *Bam*HI restriction endonuclease of the recipient *B. amyloliquefaciens* strain. Methylation *in vitro* or removal of the site should circumvent the problem.

Plasmids carrying erythromycin or chloramphenicol markers (e.g., pE194 and pC194, respectively) give low transformation frequencies in *B. amyloliquefaciens*. More efficient recipients can be obtained by isolating plasmid-cured clones from previously selected transformants. *B. subtilis* strains prone to cell lysis may not be suitable for transformation by electroporation (Bron 1990).

Comments

The protocol may be pretested with the publicly available *B. amyloliquefaciens* strain SB-I (ATCC23843; American Type Culture Collection, Rockville, MD, USA), and the plasmid pUB110 (BGSC1E6; Bacillus Genetic Stock Center, Columbus, OH, USA); the unique BamHI site of the plasmid should be methylated or removed prior to transformation (see above). A derivative of *B. amyloliquefaciens* SB-I, ALKO2100, which is a better recipient for pE194/pC194 derived plasmids, may be obtained from the author.

Pretreatment of *B. amyloliquefaciens* cells with lysozyme or pregrowth in glycine-containing medium did not enhance the transformation efficiency (Vehmaanperä 1989).

The method also enables plasmid transfer to, at least, *B. lentus*, *B. psychrophilus* and *B. sphaericus* (Bron 1990).

For integration of plasmids into the *B. amyloliquefaciens* chromosome, a two-step approach taking advantage of the thermosensitive replicon of pE194-derived plasmids is recommended (Vehmaanperä *et al.* 1991).

References

Aunstrup K, Andresen O, Falch EA, Nielsen TK (1979) Production of microbial enzymes. In: Perlman D (ed.) Microbial technology, vol. 1, 2nd edn. Academic Press, New York, pp 281-309

Bron S (1990) Plasmids. In: Harwood CR , Cutting SM (eds) Molecular biology methods for *Bacillus*. John Wiley & Sons Ltd, London, pp 75-174

Vehmaanperä J (1988) Transformation of *Bacillus amyloliquefaciens* protoplasts with plasmid DNA. FEMS Microbiol Lett 49:101-105

Vehmaanperä J (1989) Transformation of *Bacillus amyloliquefaciens* by electroporation. FEMS Microbiol Lett 61:165-170

Vehmaanperä J, Steinborn G, Hofemeister J (1991) Genetic manipulation of *Bacillus amyloliquefaciens*. J Biotechnol 19:221-240

Part III

Medical and Veterinary Applications

Electrotransformation of *Yersinia ruckeri*

JUAN M. CUTRÍN, ALICIA E. TORANZO and JUAN L. BARJA

Introduction

Electroporation may be used for transferring macromolecules (proteins and nucleic acids) into eukariotic and prokariotic cells. Presumably the high voltage induces the formation of transient pores at lipidic-protein junctions, allowing introduction of exogenous macromolecules into the cells. In the genus *Yersinia* transformation by electroporation technique was applied to species like *Y. enterocolitica, Y. pestis, Y. pseudotuberculosis* (Conchas and Carniel, 1990) and *Y. ruckeri* (Cutrín et al., 1994) One consideration must be take into account when electroporation is attempted: all the buffers must be low ionic strengh solutions because the conductivity of the sample can result in arcing in the sample chamber. Usually this conflict is avoided when bacteria is highly concentrated on the lowest ionic buffers so that they can survive (Trevors J T, 1991). The protocol presented here, and the buffers used, are optimized for *Y. ruckeri* strains.

Transformation efficiency was, in all bacterial species, dependent of the field strength. However, survival to high voltage can be dependent on the bacterial membrane. Better results were obtained with Gram-negative rather than with Gram-positive

Juan M. Cutrín, Universidad de Santiago de Compostela, Departamento de Mirobiología y Parasitología, Facultad de Biología, e Instituto de Acuicultura, Santiago de Compostela, 15706, Spain

Alicia E. Toranzo, Universidad de Santiago de Compostela, Departamento de Mirobiología y Parasitología, Facultad de Biología, e Instituto de Acuicultura, Santiago de Compostela, 15706, Spain

✉ Juan L. Barja, Universidad de Santiago de Compostela, Departamento de Mirobiología y Parasitología, Facultad de Biología, e Instituto de Acuicultura, Santiago de Compostela, 15706, Spain (*phone* +34-981-563-100(#13255); *fax* +34-981-596904; *e-mail* mpaetjlb@usc.es)

bacteria. In fact it has been reported that electroporation could cause damage in membrane, and sometimes with high voltages the pores can exceed the critical range for cell viability (Trevors et al. 1991). This procedure, compared with classical methods (calcium chloride), is efficient (about 10^5 transformants/µg DNA), simple and fast, and enables the use of frozen cell stocks. Only one disadvantage can be mentioned, the possible secondary effects on the bacteria such as stress, mutagenesis and functio-structural changes.

Materials

Equipment
- Gene Pulser^{TM} Apparatus (Bio-Rad)
- Pulse controller (Bio-Rad)
- Gene Pulser cuvettes 0.2 cm electrode gap (Bio-Rad)
- Spectrometer Lambda 2 (Perkin Elmer)
- Centrifuge Centrikon T-324 (Kontron)
- Centrifuge bottles 500 ml (Kontron)
- Microfuge tubes (Eppendorf)

Plasmids   Plasmid pSU2718 (Martínez et al. 1988)

Media
- LB agar medium
  - 10 g/l bacto tryptone
  - 5 g/l bacto yeast extract
  - 5 g/l NaCl
  - 15 g/l agar
- LB broth medium
  - 10 g/l bacto tryptone
  - 5 g/l bacto yeast extract
  - 5 g/l NaCl
- SOC medium
  - 2% bacto tryptone
  - 0.5% bacto yeast extract
  - 10 mM NaCl
  - 2.5 mM KCl
  - 10 mM $MgCl_2$
  - 10 mM $MgSO_4$
  - 20 mM glucose

Transformation buffer

- 272 mM sucrose
- 15% glycerol

▨ Procedure

Preparing competent cells

1. Using a sterile platinum wire, streak *Y. ruckeri* strain directly from a frozen stock (stored at -80°C) onto the surface of an LB agar plate. Incubate the plate for 16 hours at 25°C.

2. Tranfer two or three well-isolated colonies into 5 ml of LB broth in a 50 ml flask. Incubate the flask for 16 hours at 25°C.

3. Dilute the culture 1:50 with 250 ml of fresh LB broth in a 1 liter flask. Grow the bacteria at 25°C for 2-4 hours with gently shaking. For efficient transformation it is essential that the cells should be in exponential phase. To monitor the growth of the culture, determine the A_{600} every 30 minutes until a value ranging from 0.5 to 0.8 is achieved.

Note: All subsequent steps in the procedure should be carried out aseptically and on ice. Tubes, rotor and pipettes should be used chilled.

4. Transfer the cells to sterile ice-cold 500 ml bottles (Kontron). Cool the cultures by storing the tubes on ice.

5. Harvest the cells by centrifugation at 4,000 x g for 15 min at 4°C in a Kontron A6.9 rotor (or its equivalent).

6. Discard carefully the media from the cell pellet. Stand the bottles in an inverted position for 1 minute to allow the last traces of media to drain away.

7. Resuspend and wash the pellet in 100 ml of cold glass-ultra-filtrated water (Milli-Q, Millipore) by gently vortexing.

8. Harvest the cells by centrifugation at 4,000 x g for 15 min at 4°C in a Kontron A6.9 rotor (or its equivalent).

9. Resuspend and wash the pellet in 10 ml of ice cold transformation buffer by gently vortexing.

10. Harvest the cells by centrifugation at 4,000 x g for 15 min at 4°C in a Kontron A6.9 rotor (or its equivalent).

11. Resuspend the pellet in 0.5-1 ml of ice cold transformation buffer.

12. Working quickly dispense aliquots of the suspension into chilled sterile microfuge tubes. Depending on the requirement, 100 µl aliquots of the competent cell suspension will be more than adequate.

Note: Cell concentration can be calculated by plating on LB agar medium and it should be around 10^{10} cfu/ml.

13. Immediately snap-freeze the competent cells by immersing the closed tubes in dry ice with absolute ethanol or in liquid nitrogen.

14. Store the tubes at -80°C until needed. Aliquots also can be used directly without freezing.

Transformation protocol

1. Remove a tube of competent cells from the -80°C freezer. Thaw the cells by holding the tube in the palm of your hand. Just as the cells thaw, tranfer the tube to an ice bath.

Note: All pipette tips, pasteur pipettes, microfuges tubes, electroporation cuvettes and buffers needed after this, should be chilled.

2. Using a chilled, sterile pipette tip, tranfer 40 µl of competent cells (5 x 10^{10} cells/ml) to a chilled, sterile microfuge tube.

3. Add 1 µl of plasmid DNA (50-100 ng/µl for best results) to the same tube.

Note: Plasmids were extracted by alkaline lysis (Birnboim and Doly 1979) and purified by equilibrium centrifugation in CsCl-Ethidium bromide gradient (Sambrook et al., 1989). Ethidium bromide was removed from the collected plasmid DNA by extraction with 1-butanol, as well as CsCl by dialysis for 24-48 h against several changes of TE (pH 8.0). Plasmid DNA were precipitated with cold ethanol and resupended in Milli-Q water. The

DNA concentration of these stocks was estimated by absorbance at 260nm (assuming that one A_{260} unit is 50 µg/ml) and also by comparing the band intensity on an ethidium bromide-stained agarose gel with known concentrations of λ DNA.

4. Mix the suspension by pipetting. Leave on ice for 1 minute.

Note: Two controls should be included on the experiments; positive control may be competent bacteria which are electrotransformed with a known amount of standard preparation of DNA, and a negative control of competent bacteria without DNA.

5. Set Gene Pulser at 2.5 kV and 25 µF. Set the pulse controller at 400 Ω.

6. Tranfer mixture to cold 0.2 cm electroporation cuvette and shake suspension to the botton of the cuvette.

7. Put the cuvette in the chamber slide and apply a single electric pulse. Check voltage and time constant after pulse.

Note: τ should be around 9-9.5.

8. Immediatly after discharge, remove the cuvette from the chamber slide and add 1 ml of SOC medium. Resuspend quickly with a pasteur pipette.

9. Transfer the cells to sterile test tube and incubate with gentle shaking at 25°C for 2 hours.

10. Transfer 100 µl of transformed competent cells onto agar selective medium: LB agar containing 20 mM glucose, and an antibiotic concentration that inhibit the rest of the cells (chloramphenicol, ampicillin, kanamycin or tetracycline, depending on the plasmid used) to identify the transformants.

11. Using a sterile Drigralsky rod, spread the transformed cells over the surface of the agar plate. Leave the plates at room temperature until the liquid has been absorbed.

12. Invert the plates and incubate at 25°C. Colonies should appear in 12-16 hours.

Note: Transformed cells should be plated at low density ($<10^4$ colonies per 90-mm plate).

Results

The best transformation efficiency obtained was 6.0×10^5 transformants/µg DNA. The most important variable on the electrotransformation is the competent cells. When cells were taken from early exponential phase (absorbances of 0.5 to 0.8) the transformation efficiency ranged between 10^5 to 10^6 transformants/µg DNA, however if the cells were harvested in late exponential or stationary phase the efficiency decreased until 10^3 and 10^4 transformants/µg DNA. On the other hand with cell concentration of less than 10^8 cfu/ml, usually no transformants were obtained, however with higher concentration, the transformation efficiency is directly proportional to the number of cells (Trevors J H, 1991). Optimal results were obtained with the highest cell concentration (5.1×10^{10} cfu/ml).

Increased electrical field-strengths resulted in greater yield of transformants, however the survival rate decreased at the same time. The optimal voltage (12.5 KV/cm), which produced the highest number of transformants, is achieved with a survival rate of only 0.5%. After electrical discharge bacterial suspension remained in "dormant" phase without duplication, and needed a recuperation time of at least two hours for the expression of antibiotic resistance.

The RC (τ) time constant, a parameter that indicates the moment at which the exponential decay pulse rises to 37% of the initial field strength, determined by the capacitance (C) and the resistance (R), is an indicator of the success of the electroporation. In our experiments the best results fell on $\tau \approx 9.2$. A decrease in this value is probably due to an increase of conductivity on the electroporation mixture because of high ionic strength of the compounds that can carry temperature jumps during electrotransformation which cause potential deletorius effects on the cells.

It is noted that plasmid vectors and resident plasmids are not modified in host strains after electrotransformation. We obtained transformants with plasmids of several molecular mass ranging from 2.3 Kb to 12 Kb. With the three conformations: uncut (supercoiled), cut (linearized) and ligated (relaxed circular) forms, transformation is successful (Cutrín et al. 1994). On the other hand, using either heterologous DNA (from *E. coli*) or homologous DNA (retrotransformation), the transformation efficiencies were similar. Also, rapid plasmid DNA extractions (Birnboim and Doly method, without purification on CsCl gradient) gave as good results as those obtained with highly purified DNA.

Troubleshooting

- Time constant is lower than expected after discharge of an electric pulse:
  High ionic strength of the mixture. Be sure that your low ionic tranformation buffer is well prepared.

- Arcing when an electric pulse was applied:
  High ionic strengh on the mixture. Be sure of your low ionic tranformation buffer is well prepared.

References

Birnboim A, Doly J (1979) A rapid extraction procedure for screening recombinant plasmid DNA. Nucleic Acids Res. 7: 1513-1525.

Conchas R F, Carniel E (1990) A highly efficient transformation system for *Yersinia* species via electroporation. Gene 87: 133-137.

Cutrín J M, Conchas R F, Barja J L, Toranzo A E (1994) Electrotransformation of *Yersinia ruckeri* by plasmid DNA. Microbiología SEM 10: 69-82.

Martínez E, Bartolomé B, de la Cruz F (1988) pACY184-derived cloning vectors containing the multiple cloning site and *lacZα* reporter gene of pUC8/9 and pUC18/19 plasmids. Gene 68: 159-162.

Sambrook J, Fristsch E F, Maniatis T (1989) Molecular cloning. A laboratory manual. Cold Spring Harbor Laboratory, Cold Spring Harbor, New York.

Trevors J T (1991) Electrotransformation of bacteria. Meth. Mol. Cell. Biol. 2: 247-253.

Trevors J T, Chassy B M, Dower W J, Blaschek H P (1991) Electrotransformation of bacteria by plasmid DNA. In: Chang D C, Chassy B M, Saunders J A, Sowers A E (eds) Handbook of electroporation and electrofusion. Academic Press, New York, pp 265-289.

Electrotransformation of Enterococci

ALBRECHT MUSCHOLL-SILBERHORN and REINHARD WIRTH

Introduction

Species of the genus *Enterococcus*, first of all *Enterococcus faecalis*, are becoming more and more important as a major cause of
nosocomial infections; they can cause severe diseases of the ur-
inary tract, septicemia, and endocarditis. Another reason for the
high interest in the latter species is the unique feature of the so-
called "sex pheromone system", obviously restricted to *E. faecalis*. Other Enterococci are by far less investigated, though many of
them represent ecologically and veterinarily relevant species.

As for most other naturally not transformable Gram-positive
Bacteria, genetic analysis of Enterococci was impeded for a long
time by their strong resistance to artificial DNA-transfer. The use
of conjugative plasmids or transposons gave good results in
some aspects such as general mutagenesis but failed for cloning
of specific DNA-fragments. The central problem for an effective
transformation system was to overcome the strong cell-wall barrier without quantitatively killing the cells. In the mid-eighties
this was achieved by lysozyme-mediated generation of protoplasts in an isotonic medium and transformation in the presence
of polyethylene glycol (Wirth et al., 1986). However, the regeneration of protoplasts to viable cells was a time-consuming and
critical step. With the development of electroporation not only
protoplasted (Fiedler and Wirth, 1988) but also more or less in-

✉ Albrecht Muscholl-Silberhorn, Universität Regensburg, Institut für
Mikrobiologie, Universitätsstr.31, Regensburg, 93053, Germany
(*phone* +49-941-943-1828; *fax* +49-941943-1824;
e-mail albrecht.muscholl@biologie.uni-regensburg.de)
Reinhard Wirth, Universität Regensburg, Institut für Mikrobiologie,
Universitätsstr.31, Regensburg, 93053, Germany

tact cells became accessible to transformation (Friesenegger et al., 1991). Nevertheless, weakening of the cell-wall remained an indispensable prerequisite for reproducibly high yields of transformants. Growth of cells in the presence of high glycine concentrations turned out to be the method of choice, and different protocols have been worked out by various groups (Cruz-Rodz and Gilmore, 1990; Dunny et al., 1991). The procedure described below is based on those protocols and combines reproducibility and high yields with easy handling. The conditions have been worked out for different strains of E. faecalis, but our previous work (Friesenegger et al., 1991) suggests that some other species might behave in a similar way. It has to be noted, however, that results may vary largely even between different strains of one species, as shown in the latter reference.

Materials

- Growth medium
  - Two-fold concentrated THB-medium (Oxoid) or other medium of choice (see comments below)/1 M sucrose (autoclave as a mixture; store in the dark).
  - 20 % glycine (preheat to dissolve completely before autoclaving; store above 20 °C to avoid cristallization).
  - Sterile water
  - Mix components to result in 1x THB/ 0.5 M sucrose and a glycine concentration optimized for the strain of interest (see below).
- Washing solution: 10 % glycerol/ 0.5 M sucrose (in bidestilled water)
- Agar for plating: 1x THB/ 20 % sucrose/ 1.5 % agar (add selective antibiotics after autoclaving).

Procedure

Preparation of electrocompetent cells

1. Test your strain of choice for its glycine resistance. Glycine concentration should lead to a +/- ten-fold reduction of growth (as compared to cultures without glycine) after over-

night incubation (OD_{600} = 0.3 - 0.6). Try glycine concentrations between 2 and 8 %. Working concentration is about 1 % below minimal inhibitory concentration (inhibiting growth totally).

2. Grow cells overnight in 1x THB/ 0.5 M sucrose containing glycine in the concentration determined above.

3. Harvest cells by centrifugation (speed limitation is not a crucial point). Resuspend in prewarmed medium (same amount and composition as for overnight growth).

4. Continue incubation at 37 °C for one additional hour.

5. Harvest cells and wash with washing solution at room temperature:

   – When using large cultures wash three times with amounts of washing solution decreasing from 1 over 0.5 to 0.1 volumes of the original culture. Finally resuspend cells in 1/100 of original volume and aliquot into 40 µl portions.

   – The latter washing procedure is rather time-consuming and results either in rather great losses of cell mass or low washing efficiencies. Therefore, it is more efficient and even time-saving to aliquot cells into micro-centrifugation cups **before** washing (to the final amount corresponding to 40 µl of 100-fold concentrated cells. For convenience, harvest the total culture, decant roughly to avoid losses, resuspend in washing solution corresponding to 1/4 of culture volume, and aliquot into 1 ml portions). Wash 3 times with 1 ml each of washing solution without changing the cup. 2 - 3 minutes of high-speed centrifugation are enough for pelleting. Finally resuspend cells in 40 µl of washing solution.

   – Freeze aliquots at - 70 °C for long-time storage (up to one year) or place on ice for immediate use.

Transformation

1. Thaw aliquots on ice, mix with 5 µl of DNA-solution (max. 1 µg in bidestilled H_2O) and keep on ice for at least 5 minutes.

2. Transfer into a 0.2 cm-electroporation cuvette and immediately pulse at 1250 V/cm (we use a Gene Pulser II® with Pulse Controller plus (Biorad) at 25 µF, 2500 V, and 200 Ω. The time constant depends on the equipment used (for the mentioned unit 4.88 - 4.96 msec are usual); it should be only slightly reduced compared with cell-free washing solution.

3. If arching occurs dilute with additional 40 µl of cells and 40 µl of chilled washing solution, incubate for some minutes on ice, and try again. Arching reduces the number of transformants, but normally not so dramatically as observed for *E. coli*.

4. Immediately dilute with 200 - 1000 µl of cold 1 x THB/ 0.5 M sucrose (without glycine and antibiotics), tranfer cells to the original micro-centifugation tube, and place on ice for 5 minutes.

5. Transfer to a 37 °C water bath and incubate 1.5 - 2 h for phenotypic expression of the resistance marker(s). (Add subinhibitory amounts of antibiotic if induction is required.)

6. Spread on agar plates containing THB, 20 % sucrose and selective amounts of antibiotic(s). Colonies of transformants will be visible after 1 - 2 days of incubation.

Results

About $x * 10^5$ transformants/ µg DNA can be expected regularly from this protocol when using a well transformable strain. Results may vary considerably under identical conditions with identical probes. Therefore, if DNA is not limiting, parallel transformations should be carried out. Higher yields of up to 4×10^6 transformants/ µg DNA have been reported in the literature cited and sometimes result from the described procedure, too. It has to be noted, however, that best results are obtained by use of CsCl-purified supercoiled DNA, which is normally not available for routine applications such as molecular cloning. Small-scale plasmid preparations and especially ligation mixes generally give much lower yields. Whenever possible (i.e., no toxic effects on *E. coli* are observed), the best method remains to clone a specific DNA-fragment into *E. coli/Enterococcus*-shuttle vectors, propagate the construct in *E. coli*, and finally transform the purified recombinant product into *Enterococcus*.

Comments

- As noted above, DNA-purity is an important factor for successful transformation. Whatever purification method is used, remove all impurities, but first of all any traces of salt, as carefully as possible. Extensive washing with 70 % Ethanol after alcohol precipitation and quantitative removal of supernatant with a micropipet (rather than narrowing in a vacuum centrifuge!) often improves quality considerably. Always resuspend in bidestilled water.

- The choice of growth medium may have considerable effects on transformation efficiency, as was shown for the electroporation procedure worked out by Dunny and coworkers (1991). In this case, the number of transformants obtained by use of THB-medium was up to hundred-fold lower than, e.g., for BYGT-medium (1.9 % brain heart infusion (Difco), 0.5 % yeast extract, 0.2 % Glucose, 0.1 volume of 1 M Tris/HCl pH 8.0).

- We tried to improve electroporation efficiency by further weakening cell wall integrity. These attempts included: i) short enzymatic treatment with 10 µg/ml lysozyme, proteinase K, or pronase immediately before washing; ii) physical damage by freeze-thaw-cycles on harvested cells; and iii) two or more repeated electroporation pulses carried out on the same cell/DNA-mixture. All these treatments **per se** indeed gave better results if applied on otherwise untreated cells. Nevertheless, none of them was added to our protocol since a) their effects were **not additive** to glycine treatment, and b) they further reduced reproducibility, i.e. best results often were obtained in cases when identical experiments failed to produce transformants at all. This is obviously due to the narrow gap between maximum transformability and viability of cells.

- During preparation of electrocompetent cells all steps were carried out at room temperature. In corresponding experiments we could not see any difference in electrocompetence when compared to cells prepared on ice. In addition, it is not always simple to keep cells at a **constantly low** temperature during the whole procedure (e.g. during centrifugation).

- There still does not exist an electroporation procedure for *Enterococcus* accepted as optimum by all research groups. For example, the procedure of Dunny et al. (1991) has been worked out very well, and optimum conditions for a great number of parameters were tested systematically. However, for unknown reasons, in our hands this protocol did not work properly. The reader is encouraged to try other procedures if what is described here should not give satisfactory results for their special enterococcal strain.

Acknowledgements. We are grateful to Elisabeth Silberhorn for her technical help carried out during the optimization experiments.

References

Cruz-Rodz AL, Gilmore MS (1990) High efficiency introduction of plasmid DNA into glycine treated *Enterococcus faecalis* by electroporation. Mol Gen Genet 224:152-154

Dunny GM, Lee LN, LeBlanc DJ (1991) Improved electroporation and cloning vector system for Gram-positive bacteria. Appl Env Microbiol 57:1194-1201

Friesenegger A, Fiedler S, Devriese LA, Wirth R (1991) Genetic transformation of various species of *Enterococcus* by electroporation. FEMS Microbiology Letters 79:323-328

Fiedler S, Wirth R (1988) Transformation of bacteria with plasmid DNA by electroporation. Anal Biochem 170:38-44

Wirth R, An FY, Clewell DB (1986) Efficient protoplast transformation system for *Streptococcus faecalis* and a new *Escherichia coli-S. faecalis* shuttle vector. J Bacteriol 165:831-836

Prevotella bryantii, P. ruminicola and *Bacteroides* Strains

HARRY J. FLINT, JENNIFER C. MARTIN
and ANDREW M. THOMSON

Introduction

Representatives of the *Bacteroides/Flavobacterium/Cytophaga* phylum of Gram-negative eubacteria are normally among the most numerous inhabitants of the rumen and of the human colon. Strains formerly regarded as *Bacteroides ruminicola*, isolated from the rumen and from the pig caecum, are genetically diverse and have now been reclassified into four species belonging to the genus *Prevotella: P. ruminicola, P. bryantii, P. brevis* and *P. albensis* (Avgustin et al. 1997, Shah and Collins 1990). Based on 16rDNA sequencing, the rumen *Prevotella* spp. belong to a broad cluster that is distinct from human colonic *Bacteroides* spp. (Avgustin et al. 1994). There is evidence that at least some plasmid replicons and marker genes can function in representatives of both groups, but replicons and marker genes derived from *E. coli* plasmids are seldom if ever found to function in *Bacteroides/Prevotella* spp. (Salyers and Shoemaker 1997). Human colonic *Bacteroides* and rumen *Prevotella* species are thought to play important roles in the metabolism of starch, protein, peptides, hemicellulose and pectin while human colonic *Bacteroides* spp. also metabolise host-derived polysaccharides (Salyers 1984). The phylum also contains opportunistic pathogens such as *B. fragilis* (Salyers 1984, Shah 1992).

✉ Harry J. Flint, Rowett Research Institute, Greenburn Road, Bucksburn, Aberdeen, AB21 9SB, UK (*phone* +44-1224-716651; *fax* +44-1224-716687; *e-mail* h.flint@rri.sari.ac.uk)
Jennifer C. Martin, Rowett Research Institute, Greenburn Road, Bucksburn, Aberdeen, AB21 9SB, UK
Andrew M. Thomson, Royal Perth Hospital, University Department of Medicine, GPO Box X2213, Perth, WA6001, Australia

Most *Prevotella* and *Bacteroides* strains are sensitive to oxygen exposure which means that they must be grown in prereduced media, with an anaerobic gas phase. For strictly anaerobic rumen strains, electroporation and subsequent incubation of transformants must be conducted in an anaerobic glove box. Many of the procedural details given below refer to precautions needed to maintain anaerobiosis rather than to the electroporation conditions themselves. Human colonic *Bacteroides* strains generally show much greater tolerance of oxygen than rumen *Prevotella* strains however, and several steps (eg. electroporation, dilution series, plating) can be performed aerobically (Smith et al 1990). Rumen fluid based media are described here for the initial growth of both rumen *Prevotella* and colonic *Bacteroides* strains, but alternative media (see Holdeman et al. 1977, Smith et al. 1990) are more commonly used for colonic *Bacteroides* spp. Several species of human colonic *Bacteroides* (including *B. uniformis* and *B. vulgatus*) have been transformed by the methods described. Only two *Prevotella* strains have been used extensively in transformation experiments in this laboratory, *P. bryantii* B_14 (formerly *B. ruminicola* subsp. *brevis* B_14) and strain NCFB 2202. The latter strain is now refered to as *Bacteroides/Prevotella* 2202 since 16SrRNA profiling suggests it ressembles *Bacteroides* more closely than *Prevotella* spp.

Materials

– Anaerobic glove box (eg. Coy Laboratory Products, Ann Arbor, Michigan, including an incubator and preferably including a gas monitor for H_2 and O_2 concentrations). The following gas cylinders are needed: 100% CO_2; 100% N_2; 80% N_2/ 10% CO_2/10% H_2 (BOC or equivalent). We maintain an atmosphere of approximately 55% CO_2/40% N_2/5% H_2 by combining equal volumes of 100% CO_2 and 80% N_2/10% CO_2/ 10% H_2, although an atmosphere of 80% N_2/ 10% CO_2/ 10% H_2 has also been used. Trays of palladium catalyst inside the glove box should be reactivated regularly (once a week) to ensure removal of traces of oxygen. With flexible Coy-type glove boxes it is advisable to partially collapse and refill the glove box with gas once a week to maintain anaerobic conditions and to avoid any build up of moisture.

- Gassing hooks connected to an oxygen-free CO_2 supply (Bryant et al 1972, Holdeman et al 1977).
- Electroporator (eg. Biorad Gene Pulser, with Pulse Controller; supply of compatible 0.2cm disposable cuvettes). This can be installed inside the glove box, but it is more convenient and safer to have only the leads and cuvette holder inside the cabinet, and the controls outside.
- Supply of autoclavable, pressure resistant, sealable 15ml glass tubes (eg. Bellco Glass Inc) and 100ml crimp seal bottles (eg. Wheaton Glass Co) suitable for anaerobic media (Holdeman et al 1977).
- Liquid nitrogen storage (optional).
- M2G , M2GSC medium [*] (per 100ml)
  - 1g casitone
  - 0.25g yeast extract
  - 0.5g glucose (M2G) **or** 0.2g glucose, 0.2g cellobiose, 0.2g soluble starch (M2GSC)
  - 0.4g $NaHCO_3$
  - 30ml clarified rumen fluid (from fistulated cow, strained through muslin and centrifuged at 10,000 g to remove particulate material and autoclaved)
  - 15ml mineral salts solution 1 (stock 3g K_2HPO_4 /L)
  - 15 ml mineral salts solution 2 (stock 3g KH_2PO_4, 6g $(NH_4)_2SO_4$, 6g NaCl, 0.6g $MgSO_4.7H_2O$ 0.6g $CaCl_2$ /L)
  - 0.1ml resazurin (0.1% aqueous solution)
  - 0.1g cysteine
  - Medium heated to boiling 3 times under 100% CO_2 to drive off oxygen before addition of cysteine and $NaHCO_3$. Then dispense into Bellco glass tubes with butyl rubber stoppers and autoclave .
  - [*](modified from med2 of Hobson 1969)
- M10G agar[*]
  - 0.1g trypticase
  - 0.05g yeast extract
  - 0.5g glucose
  - 0.01ml hemin solution (20mg/ml hemin in 50% ethanol/ 0.025M NaOH)
  - 0.1ml vitamin K solution (5mg/ml in 100% ethanol)
  - 0.04g methionine

- 0.31ml VFA mixture (17ml acetic acid, 6ml propionic acid, 4ml n-butyric acid, 1ml iso-butyric acid, 1ml valeric acid, 1ml iso-valeric acid, 1ml DL-α-methyl-butyric acid)
- 3.8ml CB mineral salts (6g K_2HPO_4/L)
- 3.8ml CB mineral salts (6g KH_2PO_4, 12g $(NH_4)_2SO_4$, 12g NaCl, 1.2g $CaCl.2H_2O$, 2.5g $Mg_2SO_4.7H_2O$/L)
- 0.1ml resazurin (0.1% aqueous solution)
- 1.5g agar
- 0.2g Na_2CO_3
- 0.1g cysteine
- Adjust to pH 6.8 with 10% NaOH. Medium heated to boiling 3 times under 100% CO_2 before addition of cysteine and Na_2CO_3. Dispense into 100ml crimp-top bottles for autoclaving.
- * (modified from Caldwell and Bryant 1966)
- Anaerobic 10% glycerol : 10% glycerol prepared in Nanopure (distilled, deionised) water, heated to boiling, bubbled with 100% CO_2 and autoclaved.

Procedure

1. Place sterile petri dishes in the glove box at least 2 days before use to remove traces of oxygen from the plastic.

2. Pour plates of M10G agar medium with appropriate antibiotic additions inside the glove box (add filter sterilised antibiotic to molten medium before plate pouring).

3. Grow a 7.5ml culture of the *Prevotella*/*Bacteroides* strain overnight at 38^0C in M2GCS medium.

4. Inoculate 100ml M2G medium by syringe with 7.5ml fresh overnight culture. Grow to an OD_{650} of approximately 1.0 (mid-exponential phase).

5. Chill cultures on ice for 30 minutes, then working inside the anaerobic glove box carefully transfer into 300ml centrifuge bottles fitted with sealing cap assemblies (Sorvall cat. no. 03937, 03278) prefilled with CO_2.

6. Centrifuge cells at 10, 000g, 20 minutes ($4°C$) (Sorvall GSA rotor or equivalent). Return to the glove box, remove super-

natant and resuspend cells in 10 ml sterile, anaerobic 10% glycerol.

7. Transfer cells to a 30ml polypropylene centrifuge tube with a sealing cap assembly (Sorvall cat. no. 03530, 03535) and centrifuge at 12,000g for 10 minutes at $4°C$. Discard supernatant and resuspend cell pellet in 0.5ml anaerobic 10% glycerol and place on ice. Cells can be used directly for transformation or stored frozen. For frozen storage transfer to Sarsted screw cap tubes pregassed with CO_2, snap freeze in liquid N_2 vapour and store at $-70°C$.

8. For electroporation place a 400µl aliquot of cells prepared in 10% glycerol in a 0.2cm cuvette that has been prechilled on ice. Add DNA (1 to 5µg) and leave for 1 minute on ice.

9. Pulse, setting Gene Pulser and Pulse Controller (Biorad) to 2.5kV, 25µFD, and 800 ohms.

10. Immediately resuspend the electroporated sample in 1.6ml M2G medium and make a dilution series to 10^{-11} (D11) in eppendorf tubes containing 900µl M2G medium. Incubate dilutions at $38°C$ for 1 to 3 hours.

11. Spread 500 µl of cells from dilutions D0 to D5 onto plates of M10 agar with appropriate antibiotic selection. Simultaneously plate 100µl samples from D9 to D11 onto plates containing no antibiotic to check survival. Incubate plates in the anaerobic glove box or in anaerobic jars for 36 to 72 hours.

Results

This method has been shown to work well with *Bacteroides* spp. and with the *Bacteroides/Prevotella* strain NCFB2202 and has been used to introduce the vector plasmids pRRI207 and pRH3 containing inserts that encode polysaccharidases from *P. bryantii* or *P. ruminicola* (Thomson et al. 1992, Daniel et al 1995). It has also been used to introduce the native plasmid, pRRI4, into *P. bryantii* B_14 (Thomson and Flint 1989, Table 1). This plasmid had previously been transfered to a rifampicin resistant derivative of B_14, F101, by conjugation from strain 223/M2/7A (Flint et al. 1988). Transforming plasmid DNA was obtained from the resulting transconjugant F115, and thus came from the same genetic background as B_14 avoiding any problems with restriction/modification systems.

Table 1. Some plasmids and shuttle vectors tested for transfer into rumen *Prevotella* strains. Except for pRRI4, which was prepared from *P. bryantii* F101, results are for DNA prepared from *E. coli* DH5α. Transformants were selected using 5µg/ml tetracycline or erythromycin, as appropriate.

Vector/ plasmid	size	replicon		selectable marker gene	transformation frequency [a]					References [b]
		Bacteroides/ Prevotella	*E.coli*	(*Bact/Prev*)	Pb	P/B	Bu	Bv	Bd	
pRRI4	19.5kb	pRRI4 (*P. ruminicola* 223/M2/7)	–	tetQ	10^6	(C)[c]				(1,2,3)
pRRI207	11kb	pRRI2 (*P. ruminicola* 223/M2/7)	pHG165	ermF	–	10^3	10^4		10^5	(4)
pRH3	8.7kb	pRRI2 (*P. ruminicola* 223/M2/7)	pBluescript	tetQ	–	10^2	10^4	10^4		(5)
pKBR23.1	19.6kb	pRRI7 (*Prevotella/ Bact.* 2202)	pBR322	ermF	–	10^2	10^1		10^3	(6)
pTC-COW	13.3kb	pB8-51 (*B. eggerthii*)	pBR328	tetQ	(C)c	(C)[c]				(3,7)
pDP1	19kb	pCP1 (*B. fragilis*)	pBR322	ermF	–	10^3				(1)

[a] Pb *P. bryantii* B_14 or F101, P/B *Prevotella/Bacteroides* strain 2202, Bu *Bacteroides uniformis* 1100, Bv *Bacteroides vulgatus* 1447, Bd *Bacteroides distasonis* 419

[b] 1.Thomson, Flint 1989; 2. Shoemaker et al. 1991; 3. Flint et al. 1988; 4. Thomson et al. 1992; 5. Daniel et al. 1995; 6. Béchet et al. 1993; 7. Gardner et al. 1996

[c] Transfer by conjugation

The influence of restriction /modification systems is strongly suggested by the fact that the plasmids pRRI207 and pDP1 gave 1000 fold higher transformation frequencies for *B. uniformis* when derived from *B. uniformis* than when derived from *E. coli* (Table 2, Thomson and Flint 1989). This has also been noted by Smith et al. (1990). Possibly for this reason, electroporation has yet to be used successfully to introduce plasmids derived from *E.coli* into *P. bryantii* B_14. It may be that electroporation will prove applicable to *Prevotella* strains other than *P. bryantii* B_14 that present fewer problems due to nuclease activity. A more extensive strain survey has not been undertaken, but appears worthwhile.

Table 2. Effect of origin of transforming DNA upon transformant yield (Thomson, Flint 1989, Thomson et al. 1992)

Plasmid	Strain from which transforming DNA obtained	Frequency of em^R transformants of *B. uniformis* 1004 or 1100
pRRI207	*B. uniformis* 1004 / pRRI207	10^7
	E. coli DH5α / pRRI207	10^4
pDP1	*B. uniformis* 1004 / pDP1	10^6
	E. coli DH5α / pDP1	10^3

Troubleshooting

For previously tested strain/vector combinations, failure to obtain transformants may reflect low numbers of viable cells. This should be checked by plating cells before and after electroporation on antibiotic-free media as described above. For strict anaerobes, poor viability before electroporation will normally be due to exposure to oxygen during preparation of cell suspensions, while poor viability after electroporation indicates suboptimal electroporation conditions. Suboptimal electroporation conditions may result from excess electrolyte which can lead to arcing. This can be avoided by careful preparation of cells and by ensuring that DNA solutions are salt free and in deionised water. The quality of the deionised water appears to be important (Thomson 1990). Pulse controller settings can be varied to obtain optimal electroporation conditions.

Newly studied strains may fail to transform for many different reasons, including high nuclease activity and poor survival. If the genetic relationship of the new strain has not been established it cannot be assumed that it will support replication of vector constructs or express marker genes used successfully for other strains.

Alternative routes which have proved successful for introducing vector plasmids into *P. bryantii* involve conjugal mobilization from a *Bacteroides* donor (Shoemaker et al. 1991, Gardner et al. 1996) which may avoid problems due to restriction endonucleases if only a single DNA strand enters the recipient cell (Shoemaker and Salyers 1997).

If plates are found to be drying out too rapidly in the glove box it may be better to incubate them in anaerobic jars.

Comments

The host range of the available *Bacteroides/Prevotella* vectors and selectable markers has still to be fully established. There are suggestions that the *ermF* gene does not express in some *Prevotella* strains, including *P. bryantii* B_14 but the *tetQ* marker used in pRH3 and pTC-COW is known to express in *P. bryantii* B_14 (Gardner et al. 1996). The pB8-51 replicon from *B. eggerthii* functions in *P. bryantii* (Shoemaker et al. 1991). The pRRI2 replicon from *P. ruminicola* 223/M2/7 is known to replicate in *B. vulgatus*, *B.uniformis* and *Prevotella/Bacteroides* strain NCFB2202, but it is not yet established that it can replicate in *P. bryantii* (Table 1).

For many recipient strains it may be possible to replace M2G and M2GSC media with media (such as M10G) that do not contain rumen fluid. However a 5-7 fold greater transformant yield was obtained for *P. bryantii* B_14 when M2G medium was used for recovery of electroporated cells compared with M10G. It made no difference whether M2G or M10G medium was used for plating (Thomson 1990).

Smith et al (1990) found that transformation frequencies with *Bacteroides* recipients were improved by inclusion of 1mM $MgCl_2$ in the 10% glycerol used for electroporation.

Applications

This electroporation method is potentially suitable for introducing native plasmids or vector constructs into many oxygen-sensitive *Prevotella* and *Bacteroides* strains. Introduction of pRH3-based constructs made in *E. coli* has been successful with some recipient strains, allowing expression of polysaccharidase genes (Daniel et al. 1995). Other vectors designed for use in *Bacteroides* hosts may also be used (Thomson & Flint 1989, Smith et al. 1990, Wells & Allison 1995). Thus the method is of potential value for studies on heterologous gene expression, gene function, targetted gene disruption and strain manipulation.

Acknowledgements. We would like to acknowledge the support of the Scottish Office Agriculture Fisheries and Food Department (SOAEFD) . We are grateful to Prof. Abigail Salyers for valuable discussion and for supplying several *Bacteroides* strains.

References

Avgustin G, Wright F, Flint HJ (1994) Genetic diversity and phylogenetic relationships among strains of *Prevotella (Bacteroides) ruminicola* from the rumen. Int J Syst Bacteriol 44:246-255

Avgustin G, Wallace RJ, Flint HJ (1997) Phenotypic diversity among rumen isolates of *Prevotella ruminicola*: proposal of *Prevotella brevis* sp nov., *Prevotella bryantii* sp. nov., and *Prevotella albensis* sp.nov. and redefinition of *Prevotella ruminicola*. Int J Syst Bacteriol 47:284-288.

Béchet M, Pheulpin P, Flint HJ, Martin J, Dubourguier H-C (1993) Transfer of hybrid plasmids based on the replicon pRRI7 from *Escherichia coli* to *Bacteroides* and *Prevotella* strains. J Appl Bacteriol 74:542-548.

Bryant, MP (1972) Commentary on the Hungate technique for culture of anaerobic bacteria. Am. J. Clin. Nutr. 25:1324-1328.

Caldwell DR, Bryant MP (1966) Medium without rumen fluid for non-selective enumeration and isolation of rumen bacteria. Appl Microbiol 14:794-801.

Daniel AS, Martin J, Vanat I, Whitehead TR, Flint HJ (1995) Expression of a cloned cellulase/xylanase gene from *Prevotella ruminicola* in *Bacteroides vulgatus*, *Bacteroides uniformis* and *Prevotella ruminicola*. J Appl Bacteriol 79:417-424

Flint HJ, Thomson, AM, Bisset, J (1988) Plasmid associated transfer of tetracycline resistance in *Bacteroides ruminicola*. Appl Env Microbiol 54:855-860

Gardner RG, Russell JB, Wilson DB, Wang G-R, Shoemaker NB (1996) Use of a modified *Bacteroides/Prevotella* shuttle vector to transfer a reconstructed β-1,4-D-endoglucanase gene into *Bacteroides uniformis* and *Prevotella ruminicola*. Appl Env Microbiol 62:196-202

Salyers AA (1984) *Bacteroides* of the human lower intestinal tract. Ann Rev Microbiol 38:293-313

Hobson PN (1969) Rumen bacteria. Methods Microbiol. 3B:133-149

Holdeman LV, Cato EP, Moore WEC (1977) Anaerobe Laboratory Manual. Fourth edition. Virginia Polytechnic Institute.

Salyers AA, Shoemaker NB (1997) Genetics of human colonic *Bacteroides*. In Gastrointestinal Microbiology vol 2 ed. Mackie RI, White BR, Isaacson RE, pp 299-320 Chapman and Hall.

Shah HN (1992) The genus Bacteroides and related taxa. In The Prokaryotes: A Handbook on the Biology of Bacteria, Isolation, Identification, Application 2nd ed., Balows A, Trüper HG, Dworkin M, Harder W and Schleifer KH chap.194 Springer Verlag New York

Shah HN, Collins, MD (1990) *Prevotella*, a new genus to include *Bacteroides melaninogenicus* and related species formerly classified in the genus *Bacteroides*. Int J Syst Bacteriol 40:205-208

Shoemaker NB, Anderson KA, Smithson SL, Wang GR, Salyers AA (1991) Conjugal transfer of a shuttle vector from the human colonic anaerobe *Bacteroides uniformis* to the ruminal anaerobe *Prevotella* (formerly *Bacteroides*) *ruminicola* B₁4. Appl Env Microbiol 57:2114-2121

Smith CJ, Parker CJ, Rogers, M (1990) Plasmid transformation of *Bacteroides* spp. by electroporation. Plasmid 24, 100-109.

Stewart CS, Flint HJ, Bryant MP (1997) The rumen bacteria. In The Rumen Microbial Ecosystem, ed. Hobson PN, Stewart CS, pp.10-72. Blackie.

Thomson AM (1990) Gene transfer in rumen *Bacteroides* species. PhD thesis, U. Aberdeen UK

Thomson AM, Flint HJ (1989) Electroporation induced transformation of *Bacteroides ruminicola* and *Bacteroides uniformis* by plasmid DNA. FEMS Microbiol Letts 61:101-104

Thomson AM, Flint HJ, Béchet M, Martin J, Dubourguier H-C (1992) A new *E.coli: Bacteroides* shuttle vector pRRI207 based on the *Bacteroides ruminicola* plasmid replicon pRRI2. Curr Microbiol 24:49-54

Wells JM, Allison A (1995) Molecular genetics of intestinal anaerobes. In Human colonic bacteria: role in nutrition, physiology and pathology. ed Gibson GR and MacFarlane GT pp.25-60. CRC Press.

Electrotransformation of *Bordetella*

GAVIN R. ZEALEY and REZA K. YACOOB

Introduction

Whooping cough (pertussis) is a disease of the upper respiratory tract caused by a localized infection by *B. pertussis*. The absence of a technique for high-frequency transformation has limited the genetic manipulation of this organism. Weiss and Falkow (1982) described that *B. pertussis* cells cannot be transformed by classical methods such as calcium chloride treatment but that a cold shock or freezing of the cells results in transformation by plasmids of the P and W incompatibility groups. The transformation frequency obtained was less than $10^3/\mu g$ of DNA however and a restriction system in *B. pertussis* prevents the introduction of plasmids containing the Hind III recognition sequence. A modified form of this technique has been used to transform *B. bronchiseptica* at a frequency of $10^4/\mu g$ of DNA (Lax, 1987). Exposure of living cells to brief pulses of electric current results in a reversible permeabilisation of the biomembranes and has been termed electroporation. This transient membrane permeability then allows DNA uptake by the cells. We describe here the applicability of electroporation to the transformation of *Bordetella pertussis* and *B. parapertussis*. Electrotransformation routinely yields $> 10^6$ transformants per μg of plasmid DNA.

✉ Gavin R. Zealey, Aventis Pasteur, 1755 Steeles Avenue West, Toronto, Ontario, M2R 3T4, Canada (*phone* +416-667-2854; *fax* 416-667-2459; *e-mail* gzealey@ca.pmc-vacc.com)
Reza K. Yacoob, Aventis Pasteur, 1755 Steeles Avenue West, Toronto, Ontario, M2R 3T4, Canada

Outline

The procedure is briefly outlined below.

1. Grow strain of *Bordetella* to be transformed in appropriate growth medium.

2. Harvest, wash and concentrate cells.

3. Resuspend cells in glycerol, aliquot and store at -70 °C.

4. Add DNA to electrocompetent Bordetella cells.

5. Subject mixture of cells + DNA to electrical pulse.

6. Add appropriate growth medium to electroporated mixture.

7. Allow cells to recover at 37 °C for 1 hour.

8. Harvest electroporated cells and plate onto selective agar media.

9. Incubate for 3-5 days.

10. Transformants.

Materials

- Electroporator (ie BTX transfector)
- 10% glycerol (BRL redistilled)
- Distilled water (or deionized)
- Bordet Gengou medium (BBL)
- Liquid pertussis growth medium

Pertussis growth medium

	(g/l)
L-proline	5
NaCl	2.5
KH2PO4	0.5
KCl	0.2
MgCl2.6H2O	0.1
Tris	1.5
Casamino acids	10.0
pH to 7.6	
Then add CaCl2.2H2O	0.02

Autoclave, then add 20 ml 50% Monosodium glutamate (filter sterilized), 10 ml 100x growth factors (filter sterilized).

Store at 4 °C.

100x Growth factors /100 ml

L-cysteine	400 mg
FeSO4.7H2O	100 mg
Niacin	40 mg
Glutathione	1.5g
Ascorbic Acid	4 g
Filter sterilize and store at 4 °C.	

▨ Procedure

Growth of Cells

1. Inoculate 10 ml of liquid pertussis growth medium (ACP) with one loopful from a fresh agar (Bordet Genou) strain of the *B. pertussis* to be transformed. Grow for 48 hr (36 °C with shaking 100 rpm).

2. Transfer 5 ml of the culture to 500 ml of fresh ACP media and grow to A_{600}=1-3. (For best transformation efficiency, this should be approximately $1x10^8$ - $1x10^9$ cells per ml).

3. Harvest cells by centrifugation (8000 g, 20 minutes 4 °C).

4. Wash the cell pellets in 500 ml of cold (4 C) distilled water. Resuspend and harvest as step 3.

5. Repeat step 4. Electrotransformation at high voltages requires cell suspensions of very low conductivity, therefore thorough washing of cells is required.

6. Wash the cell pellets in 50 ml 10% glycerol (redistilled). Resuspend and harvest as in step 3.

7. Resuspend the cell pellets in 5ml 10% glycerol.

8. Aliquot into 1 ml samples in eppendorf tubes and freeze at -70 °C. The final cell concentration should be approximately $1x10^{10}$ cells / ml. These cells can be frozen and used for up to one year.

This method should yield enough electrocompetent cells from 500 ml cultures for between 25-50 electrotransformations.

1. Thaw vials of cells on ice.

Transformation

2. Add between 100-200 µl of cells to the transforming DNA (ng-µg) in a sterile electroporation cuvette that has been pre-chilled for 5 minutes on ice. It is important to avoid bubbles in the cell suspension as these may cause arcing in the cuvette. The DNA should be as pure as possible (ie ethanol precipitated and resuspended in distilled water) and at a concentration of approximately 1µg/µl.

3. Chill the DNA + cell suspension mixture on ice for 10 min.

4. Carefully place electrode into the cuvette containing the mixture, again avoiding the introduction of bubbles. Electroporate with BTX Transfector 100 (Amplitude = 650V with power plus, Pulse length = 5 ms., Gap = 0.8 mm). This will deliver a pulse of about 25kV/cm.

5. Remove the electrode and immediately add 1ml of liquid growth medium to the cuvette and resuspend the cells.

6. Incubate the cells at 37°C with shaking (100 rpm) for 1 hour.

7. Harvest cells by centrifugation in a microfuge for 2 min, resuspend and plate onto Bordet Gengou agar with the appropriate selection media.

8. Incubate the plates for 5-6 days.

Results

Factors affecting transformation efficiency:

- No transformants were obtained if DNA was omitted from the electroporation mixture or if the cells were not subjected to an electrical discharge.

- Transformation efficiency was found to be essentially independent of cell concentration over the range of 10^8 to 10^{10} cells/ml and showed a linear response to DNA concentrations between 1 ng and 1 µg/µl. In contrast to this, transformation efficiency was found to be directly proportional to the initial electric field strength (data not shown). Thus few transformants (1-10/µgDNA) were obtained at field strengths below 8 kV/cm and the greatest efficiency ($> 10^6$/µg DNA) was obtained at about 25 kV/cm. This value is the highest field strength that can be generated by the BTX 100 Transfector under these experimental conditions. Transformation frequency was therefore determined after subjecting the cells to between 1 and 5 sequential discharges at an initial field strength of 25 kV/cm. Multiple discharges did not result in an increase in the number of transformants obtained.

- The conditions for optimal transformation of B. parapertussis were also investigated and were found to be identical to those for B. pertussis. Weiss and Falkow (1982) reported a restriction system in B. pertussis Tohama III strain that limits the introduction of plasmids containing the HindIII recognition sequence. The Connaught B. pertussis production strain appears to have a similar restriction system as shown by the reduced transformation efficiency by plasmid pRK404 which has a single HindIII site. This reduction presumably indicates that while many of the plasmid molecules introduced are degraded, some are modified by the host and are therefore maintained in the cell.

- We have used this method successfully not only with plasmid DNA (Zealey et al 1988) but have also electrotransformed *B. pertussis* with linear DNA targetted to specific loci in the chromosome by homologous recombination (Zealey et al 1990, 1992, and Loosmore et al 1995). This linear transformation has been used to specifically modify virulence genes and to manipulate gene expression in Bordetella (Loosmore et al 1995). With linear DNA the frequency of transformation drops to approximately two per μg of DNA.

Comments

Electrotransformation involves working with high voltages and high current discharges. All safety precautions indicated by the equipment manufacturer should be followed. Explosive arcing may also occur, causing aerosolization of the sample. Therefore it is best to perform electrotransformations in a bio-containment hood and adopt all precautions normally taken when working with pathogenic bacteria.

References

1. Weiss AA, Falkow S (1982) Plasmid transfer to *Bordetella pertussis*: conjugation and transformation. J Bacteriol 152:549-552
2. Lax AJ (1987) Improved *Bordetella* transformation. Nucleic Acids Res 15:856
3. Zealey GZ, Dion M, Loosmore SM, Yacoob RK, Klein MH (1988) High frequency transformation of *Bordetella pertussis*. FEMS Microbiology Letters 56:123-126
4. Zealey GZ, Loosmore SM, Yacoob RK, Cockle SA, Boux LJ, Miller LD, Klein MH (1990). Gene Replacement in *Bordetella pertussis* by transformation with linear DNA. Bio/Technology 8:1025-1029
5. Zealey GR, Loosmore SM, Yacoob RK, Cockle SA, Herbert AB, Miller LD, Mackay NJ, Klein MH (1992) Construction of *Bordetella pertussis* Strains That Overproduce Genetically Inactivated Pertussis Toxin. Appl Environ Microbiol 58:208-214
6. Loosmore SM, Yacoob RK, Zealey GR, Jackson GED, Yang Y-P, Chong PS-C, Shortreed JM, Coleman DC, Cunningham JD, Gisonni L, Klein MH (1995) Hybrid genes over-express pertactin from *Bordetella pertussis*. Vaccine 13:571-580

Suppliers

BTX
3742 Jewell Street
San Diego, CA 92109
Tel (619) 270-0861
Fax (619) 483-3817

GIBCO BRL
Grand Island. New York, USA
Tel (800) 828-6686
Fax (800) 331-2286

BBL Becton Dickinson
243 Cockeysville MD, USA
Tel (888) 237-2762
Fax (800) 847-2220

Transformation of *Campylobacter jejuni*

BEN N. FRY, MARC M.S.M. WÖSTEN, TRUDY M. WASSENAAR
and BERNARD A.M. VAN DER ZEIJST

Introduction

Campylobacter jejuni is a Gram-negative bacterium and is a
commensal of many animal species (Skirrow and Blaser,
1992). In humans it is the major cause of human bacterial enteritis both in developed and developing countries (Tauxe, 1992;
Taylor, 1992). Many cases of *Campylobacter* enteritis in humans
have been associated with the consumption of or contact with
undercooked chicken meat (Deming et al., 1987; Harris et al.,
1986).

A better understanding of the pathophysiology of the organism, its immunogenic properties, and its colonisation requirements in chickens, can lead to better treatment, vaccine development and prevention of infection, respectively. Research in
these areas is greatly dependent on genetic approaches, and
transformation is an essential tool for this.

The introduction of foreign DNA into *C. jejuni* was first described by Labigne-Roussel et al. (1987) using conjugation. Later

✉ Ben N. Fry, Royal Melbourne Institute of Technology University,
Department of Applied Biology and Biotechnology, GPO Box 2476V,
Melbourne, VIC 3001, Australia (*phone* +61-3-9660-3407;
fax +61-3-9662-3421; *e-mail* ben.fry@rmit.edu.au)
Marc M.S.M. Wösten, Utrecht University, Faculty of Veterinary Medicine,
Department of Bacteriology, Institute of Infectious Diseases and Immu-
nology, Yalelaan 1, Utrecht, 3584 CL, The Netherlands
Trudy M. Wassenaar, Johannes Gutenberg Universität, Institut für
Medizinische Mikrobiologie, Hochhaus am Augustusplatz, Mainz, 55101,
Germany
Bernard A.M. van der Zeijst, Utrecht University, Faculty of Veterinary
Medicine, Department of Bacteriology, Institute of Infectious Diseases and
Immunology, Yalelaan 1, Utrecht, 3584 CL, The Netherlands

also electro-transformation (Miller et al., 1988) and natural-transformation (Wang and Taylor, 1990) were described as methods to transform *C. jejuni*. These methods are now widely used. Since electro-transformation results in a relatively high efficiency, is simple and fast, it is the method of choice. However, electro-transformation is not always successful, and electro-competence has been shown to be strain dependent (Wassenaar et al., 1993). Therefore this chapter is not restricted to electro-transformation, but also describes the protocols for conjugation and natural-transformation as alternative methods. For a comparison study to determine the optimal conditions and to evaluate electro-transformation and natural-transformation we refer to previous work (Wassenaar et al., 1993).

Materials

Strains *Campylobacter jejuni* strains tested:
81116 (NCTC 11828; Newell et al., 1985); 480 (NTCT 12744; King et al., 1991); 11271 and 11279 (H. Lior, Laboratory Centre for Disease Control, Ottowa, Canada); 51180 and BA63923 (D.G. Newell, Central Veterinary Laboratory, New Haw, UK); 606, 719 and 756 (H. Endtz, University Hospital, Leiden, The Netherlands); 207252, 206710, 129108, 132960, 210388, 205223, 105713, 201191, 205224, 207251, 209071, 209755 and 206470 (Endtz et al., 1993).

Electrotrans- – Strains of choice: 480, 129108
formation – Skirrow agar plates (Skirrow, 1977)
– Glycerol/sucrose
  – 15% glycerol
  – 272 mM sucrose
  – sterilize by filtration
– Bio-Rad Gene Pulser with pulse controller
– Electro-transformation cuvettes: 0.56 mm (Biotechn. & Exp. Res. Inc., San Diego, CA, USA) or, alternatively, 2 mm (Bio-Rad)
– Heart Infusion broth (HI; Difco)
– Heart Infusion agar plates (2% agar)

- Strain *C. jejuni* 81116 mutant T2   Conjugation
- Skirrow agar plates (Skirrow, 1977)
- Tetracycline (Tc; 20 µg/ml; stock: 20 mg/ml in 70% EtOH)
- *E. coli* S17.1 (Simon et al., 1983)
- Luria-Broth (Sambrook et al., 1989)
- Trimethoprim (TMP; 5 µg/ml; stock: 5 mg in 1 ml MeOH)
- sterile H_2O
- Muller Hinton plates (MH; BBL)
- Heart Infusion broth (HI; Difco)
- (Cephalothin (CEP; 15 µg/ml; stock: 15 mg in H_2O))

- Strain of choice: 81116   Natural-trans-
- Skirrow agar plates (Skirrow, 1977)   formation
- Heart Infusion broth (HI; Difco)
- Heart Infusion agar (2% agar)

Procedure

Culturing *Campylobacter jejuni*

All *C. jejuni* strains were grown under micro-aerophilic conditions (5% O_2, 10% CO_2 and 85% N_2) on Skirrow agar medium at 42°C.

Electro-transformation of *Campylobacter jejuni*

Electro-transformation can be used to introduce chromosomal DNA and suicide and shuttle vectors.

1. Culture *C. jejuni* on four Skirrow agar plates overnight at   Preparation of
   42°C under micro-aerophilic conditions.   electrocompe-
                                                                       tent cells
2. Cool cells under micro-aerophilic conditions for 15' at 4°C.

3. Harvest the cells by scraping the growth off the plates with an inoculation loop and resuspend (by gently pipetting) in 5 ml ice-cold glycerol/sucrose. (An alternative method of harvesting by overlaying the agar plate with the solution and pipetting off the suspension results in a higher salt concentration, which as a consequence requires extensive washing)

4. Spin the suspension for 20' at 3000 x g, 4°C, and wash twice with 5 ml ice-cold glycerol/sucrose.

5. Resuspend cells in approximately one volume of cell pellet glycerol/sucrose ($\pm$ 500 µl). This results in approximately 10^{11} cells/ml.

6. Use immediately or freeze aliquots of 50 or 200 µl at -80°C.

Electrotrans-
formation

1. Thaw electro-competent *C. jejuni* cells slowly or use freshly prepared cells and keep on ice.

2. Add 1 µg DNA (suspended in 2O) to 50 µl electro-competent cells.

3a. Pulse in a cold **0.56** mm cuvet: **0.7 kV, 25 µF, 600 Ω**. Time constant ranges from 4 to 10.

3b. Pulse in a cold **2 mm cuvet: 2.48 kV, 25 µF, 600 Ω**. Time constant ranges from 4 to 15.

4. Flush the content of the cuvet with 1 ml HI broth and transfer to an eppendorf tube and pierce a hole in the top. Alternatively, the content of the cuvet can be flushed onto a HI agar plate.

5. Let the cells recover for 5 h at 37°C under micro-aerophilic conditions.

6. When recovering the cells on HI agar plates, harvest cells with 1 ml HI and transfer to an eppendorf tube. Otherwise go to step 7.

7. Use 10 µl to make a 10^{-5} and 10^{-6} dilution in HI for viability counts and spread onto a Skirrow plate.

8. 8. Plate 100 µl of the rest of the suspension out on Skirrow plates supplemented with the appropriate antibiotic for transformant selection.

    – When a low efficiency of transformation is expected concentrate first by centrifuging for 8' at 3000 x g and resuspend pellet in 100 µl HI before plating out.

    – When a high efficiency of transformation is expected, 10-fold and 100-fold dilutions should also be plated out to ensure single colony growth.

9. Incubate for 2 days at 42°C or 3 days at 37°C under micro-
    aerophilic conditions.

Conjugation from *E. coli* S17.1 to *Campylobacter jejuni*

Conjugation can be used to transfer a plasmid containing an origin of transfer (oriT) from *E. coli* to *C. jejuni*. The *E. coli* donor
strain has to harbour the P incompatibility group conjugative
plasmid or the genes required for mobilization. *E. coli* SM10 (Simon *et al.*, 1983) and *E. coli* S17.1 (Simon et al., 1983) are the most
commonly used donors. Conjugation has been used to transform *C. jejuni* with shuttle vectors and suicide vectors. The bot-
tleneck in this experiment is often the selection of the *C. jejuni*
from the *E.coli*. We have found that by using an antibiotic resistant *C. jejuni* strain this procedure could be facilitated. Here we
will use a genetically modified mutant of *C. jejuni* strain 81116,
T2, which is made tetracyclin (TET) resistant (Wassenaar et al.,
1995).

Grow *C. jejuni* T2 from -80°C on Skirrow plates with TET over-   **Day one**
night at 42°C under micro-aerophilic conditions.

1. Inoculate two Skirrow plates complemented with TET using   **Day two**
    the freshly grown bacteria in step A. and grow overnight at
    42°C under micro-aerophilic conditions.

2. Grow *E. coli* S17.1 with a plasmid containing an oriT over-
    night in 10 ml LB with TMP and appropriate antibiotics
    for maintenance of the plasmid at 37°C.

1. Dilute *E. coli* S17.1 cells 1:100 in LB and grow ± 2.5 hrs to a   **Day three**
    concentration of 10^8 bacteria/ml (OD_{550} = 0.5).

2. Cool *C. jejuni* cells micro-aerophilic for 15' at 4°C and harvest
    from plates by scraping with an inoculation loop and resus-
    pend in 5 ml ice-cold H_2O.

3. Per conjugation reaction, use 0.5 ml *C. jejuni* suspension,
    transfer to a 50 ml tube and fill the tube with ice-cold
    H_2O. Centrifuge 10' 3000 x g at 4°C.

4. Resuspend the *C. jejuni* pellet in 1 ml *E. coli* culture from step
    1. Overlay a MH agar plate with this suspension.

5. Incubate the agar plate for 5 hrs at 37°C, under micro-aerophilic conditions.

6. Harvest the bacteria from the plate with 1 ml HI.

7. Plate out 100 µl of this suspension on Skirrow plates with plasmid encoded antibiotic and TET for selection of *C. jejuni* transformants.

Additionally Cephalothin can be used in the Skirrow plates to prevent contamination and *E. coli* overgrowth.

Natural-transformation of *Campylobacter jejuni*

Natural-transformation is an easy way to transform *C. jejuni* strains. However, not all strains are naturally transformable (Wassenaar et al., 1993).

Day one 1. Grow *C. jejuni* from -80°C stocks under micro-aerophilic conditions on Skirrow plates overnight at 42°C.

Day two 1. Inoculate a Skirrow plate from the overnight culture and grow under micro-aerophilic conditions at 42°C for 16 hours. (Do not grow for longer than 16 hours. It is important to use fresh cells!)

2. Fill 1.5 ml tubes with 1 ml melted HI-agar and allow to solidify.

Day three 1. Harvest cells from one plate by scraping with an inoculation loop and resuspend in 1 ml HI.

2. Layer 200 µl cell suspension onto the HI agar in the tubes of step B2. Close lids and pierce them.

3. Incubate under micro-aerophilic conditions at 37°C for three hours.

4. Add DNA to the top layer (i.e. the cell suspension):
   - Suicide Vectors: 10 µg
   - Shuttle Vectors: 0.1 - 1 µg (varies with plasmid type)
   - Chromosomal DNA: 0.1 - 1 µg
     Mix cells with DNA by pipetting up and down.

Note: Do not leave cells exposed to oxygen longer than necessary!

5. Incubate under micro-aerophilic conditions at 37°C for three hours.

6. Resuspend cells and transfer the total volume to Skirrow plates supplemented with the plasmid encoded antibiotic for selection. (Use fresh plates. Make sure the plates are not too wet! Dry first.)

Incubate under micro-aerophilic conditions at 42°C for two days or at 37°C for three days.

Results

Some results obtained with the three different procedures are given below. An indication of the efficiencies is given for a number of strains. The transformation of shuttle vectors, suicide vectors and chromosomal DNA are given in separate tables.

Shuttle vector DNA

Strains 11279, 51180, BA63923, 606, 719, 756 could not be transformed by electro- and natural- transformation. Conjugation was not tried for these strains.

Strains 207252, 206710, 132960, 210388, 205223, 105713, 201191, 205224, 207251, 209071, 209755 could not be transformed by electro-transformation. Conjugation and natural-transformation was not tried for these strains.

Table 1. *C. jejuni* colony forming units (CFU's) resistant to the antibiotic kanamycin after transfer of the shuttle vector pILL550 (Labigne-Roussel et al., 1987) isolated from *E. coli*.

C. jejuni strain	Electro-transformation	Conjugation	Natural-transformation
81116	0	5-50	0
81116[a]	5000	na	0
81116[b]	nd	na	50,000-100,000

Table 1. Continous

C. jejuni strain	Electro-transformation	Conjugation	Natural-transformation
81116[c]	nd	na	40-800
480	10,000-50,000	nd	0
480[a]	5,000-40,000	na	0
480[d]	10,000-50,000	na	na
11271	10-250	nd	0
129108	400-600	nd	0
129108[a]	40,000-60,000	na	0
206470	25-35	nd	nd

[a] Using a shuttle vector isolated from the host cell.

[b] Using a shuttle vector isolated from the host strain containing host strain DNA.

[c] Using a shuttle vector isolated from the host cell and using a host cell that harbours a vector containing homologous sequences to the donor vector

[d] Using fresh cells instead of cells that had been stored at -80°C.
nd, not determined.
na, not applicable.

Suicide vector DNA

Table 2. *C. jejuni* colony forming units (CFU's) resistant to the antibiotic kanamycin after transfer of a suicide vector isolated from *E. coli* containing host cell DNA interrupted by a kanamycin resistance cassette.

C. jejuni strain	Electro-transformation	Conjugation	Natural-transformation
81116	1-5	1-4	10-200
129108	10	nd	0

nd, not determined

Chromosomal DNA

Table 3. *C. jejuni* colony forming units (CFU's) after transfer of chromosomal host cell DNA.

C. jejuni strain	Electro-transformation	Conjugation	Natural-transformation
81116	>100,000	na	›100,000
480	>100,000	na	0
480[a]	0	na	na

[a] Omission of pulse when using the electro-transformation procedure.
na, not applicable

Applications

The choice of which method to use is determined by the competence of the *C. jejuni* strain of choice, and also depends on the source and nature of the DNA that needs to be transformed.

When using a shuttle vector, we found that when the vector was isolated from the donor strain the efficiency was higher than when it was isolated from another strain or organism. This is thought to be due to a restriction modification system in *C. jejuni* strains. The observation that most tested strains can not be transformed could also be due to this phenomenon.

For the most extensively studied strain, 81116, the presence of some host cell DNA on the plasmid greatly enhances the transformation efficiency and enables natural-transformation.

Introducing suicide vectors in this strain is most efficiently done by natural-transformation.

The transformants obtained when using the electro-transformation protocol could also have resulted from natural transformation. For instance, the transfer of homologous chromosomal DNA was very efficient in the two strains tested (Table 3). However, since strain 81116 is naturally transformable the results obtained with electro-transformation could be due to the uptake of the DNA by the mechanism of natural-transformation. Strain 480 is not naturally competent therefore, DNA transfer is exclusively due to electro-transformation.

References

Deming, M.S., Tauxe, R.V., Blake, P.A., Dixon, S.E., Fowler, B.S., Jones, T.S., Lockamy, E.A., Patton, C.M., and Sikes, R.O. (1987) *Campylobacter* enteritis at a university: transmission from eating chicken and from cats. Am J Epidemiol 126: 526-34.

Endtz, H. P., Giesendorf, B. A., van Belkum, A., Ragsdale, C. W. and Quint, W. G. 1993. PCR-mediated DNA typing of *Campylobacter jejuni* isolated from patients with recurrent infections. Res. Microbiol. 144:703-708

Harris, N.V., Weiss, N.S., and Nolan, C.M. (1986) The role of poultry and meats in the etiology of *Campylobacter jejuni/coli* enteritis. Am J Public Health 76: 407-11.

King, V., Wassenaar, T.M., Newell, D.G., and Van der Zeijst, B.A.M. (1991) Variations in *Campylobacter jejuni* flagellin genes, during in vivo and in vitro passage. Micr Ecol Health Dis 4: 135-140.

Labigne-Roussel, A., Harel, J., and Tompkins, L. (1987) Gene transfer from *Escherichia coli* to *Campylobacter* species: development of shuttle vectors for genetic analysis of *Campylobacter jejuni*. J Bacteriol 169: 5320-5323.

Miller, J.F., Dower, W.J., and Tompkins, L.S. (1988) High-voltage electroporation of bacteria: genetic transformation of *Campylobacter jejuni* with plasmid DNA. Proc Natl Acad Sci U S A 85: 856-860.

Newell, D.G., McBride, H., and Dolby, J.M. (1985) Investigations on the role of flagella in the colonization of infant mice with *Campylobacter jejuni* and attachment of *Campylobacter jejuni* to human epithelial cell lines. J Hyg (Lond) 95: 217-227.

Sambrook, J., Fritsch, E.F., and Maniatis, T. (1989) Molecular Cloning: A laboratory Manual. Cold Spring Harbor Laboratory, 2nd ed. Cold spring Harbor, New York.

Simon, R., Priefer, U. and Puhler, A. (1983) A broad host range mobilization system for *in vivo* engineering: transposon mutagenesis in gram negative bacteria. Bio/Technology 1, 784-791.

Skirrow, M.B. (1977) Campylobacter enteritis: a "new" disease. Br Med J 2: 9-11.

Skirrow, M.B., and Blaser, M.J. (1992) Clinical and epidemiologic considerations. In Campylobacter jejuni: Current Status and Future Trends. Nachamkin, I., Blaser, M. J. and Tompkins, L. S. (eds). Washington, DC: American Society for Microbiology, pp. 3-8.

Tauxe, R.V. (1992) Epidemiology of *Campylobacter jejuni* infections in the United States and other industrialized nations. In: Campylobacter jejuni: Current Status and Future Trends. I. Nachamkin, M. J. Blaser & L. S. Tompkins (eds). Washington, DC: American Society for Microbiology, pp. 9-19.

Taylor, D.N. (1992) *Campylobacter* infections in developing countries. In Campylobacter jejuni: Current Status and Future Trends. I. Nachamkin, M. J. Blaser & L. S. Tompkins (eds). Washington, DC: American Society for Microbiology, pp. 20-30.

Wang, Y., and Taylor, D.E. (1990) Natural transformation in *Campylobacter* species. J Bacteriol 172: 949-55.

Wassenaar, T.M., Fry, B.N.,and Van der Zeijst, B.A.M. (1993) Genetic manipulation of *Campylobacter*: evaluation of natural transformation and electro-transformation. Gene 132:131-135

Wassenaar, T.M., Fry, B.N. and van der Zeijst, B.A.M. (1995) Variation of the flagellin gene locus of *Campylobacter jejuni* by recombination and horizontal gene transfer. Microbiology 141: 95-101.

Slow-Growing Mycobacteria

BARRY J. WARDS and DESMOND M. COLLINS

Introduction

Mycobacteria are gram-positive organisms and are divided into fast- and slow-growing species. The latter group contains the major human and animal pathogens, *Mycobacterium tuberculosis* and *Mycobacterium bovis* as well as *Mycobacterium paratuberculosis* and members of the *Mycobacterium avium* complex. Their characteristics include complex cell walls, slow growth, clumping and inefficient genetic transfer systems. These characteristics have placed significant limitations on genetic manipulation of these species. The development of efficient transformation methods using electroporation has significantly improved and expanded the types of genetic approaches available for analysing mycobacterial genes. In particular, more efficient transformation efficiencies have improved the likelihood of:

- Reliably producing recombinants

- Obtaining high quality representative libraries

- Performing homologous recombination strategies with suicide plasmids.

✉ Barry J. Wards, Wallaceville Animal Research Centre, New Zealand Pastoral Agriculture Research Institute Limited, PO Box 40063, Upper Hutt, New Zealand (*phone* +64-4-5286089;
fax +64-4-5281380; *e-mail* wardsb@agresearch.cri.nz)
Desmond M. Collins, Wallaceville Animal Research Centre, New Zealand Pastoral Agriculture Research Institute Limited, PO Box 40063, Upper Hutt, New Zealand

▨ Materials

- Middlebrook 7H9 broth (Difco) and Middlebrook 7H11 agar   **Media**
  (Difco). Both supplemented with 10% v/v albumin/glucose
  complex (Difco), 0.05% v/v Tween 80, 0.4% w/v sodium pyr-
  uvate (for *M.bovis*) or 0.5% v/v glycerol (for *M.tuberculosis*).
  Antibiotics are added to solid medium as appropriate.
- Antibiotics for transformant selection. Kanamycin sulphate
  (Sigma) is a commonly used antibiotic selection for mycobacterial vectors. Prepare as a 10 mg/ml stock (filter-steri-
  lize); store at -20°C. Use at 20 µg/ml.
- 10% v/v glycerol
- 30% w/v glycine

- Electroporation apparatus with pulse controller (e.g., Gene   **Equipment**
  Pulser, Bio-Rad)
- Electroporation cuvettes: 0.2 cm-gap electrodes (e.g., Gene
  Pulser/E.coli Pulser Cuvettes, Bio-Rad)
- Roller bottle apparatus

▨ Procedure

Note: *M.tuberculosis* and *M.bovis* are Class II pathogens. All ex-   **Biosafety:**
perimental procedures must be conducted using suitable protec-   **mycobacteria**
tion in a Class II Biosafety Cabinet under appropriate containment conditions.

1. Prepare a starter culture by inoculating a single colony (or   **Mycobacterial**
   inoculum from a frozen stock) into 5 ml 7H9 broth. Incubate   **culture**
   at 37°C without shaking until cells are growing logarithmi-
   cally (~1-2 weeks).

2. Inoculate 300-500 µl of the starter culture into 100 ml of pre-
   warmed 7H9 broth in 1 litre bottles (Schott). Incubate at 37°C
   on a roller bottle apparatus at 1-1.5 rev/min.

Note: Slow-growing mycobacteria do not grow well with vigorous shaking. Improved growth rate can be achieved with mild
aeration provided by the roller apparatus. The slow turning
also reduces cell clumping and keeps the culture homogeneous.

3. Continue incubation until the culture has reached an OD_{600} of ~1. This generally takes 7-9 days.

Note: As a general guide, an OD_{600} of ~0.6 is reached when a finger is **just visible** through the culture when holding the bottle up to the light. An OD_{600} of ~1 should be reached in another 2-3 days. The culture should appear smooth and unclumped.

4. Optional - Transformation efficiency can be improved 2-4-fold through addition of glycine(1.5% for 10 h) or ethionamide (2 µg/ml for 5 h) to the culture prior to harvest.

Note: After 10h incubation with glycine the cultures will often have a congealed appearance. However, a homogeneous culture should be obtained after gentle mixing.

Cell preparation

5. Harvest the cells in sealed buckets and tubes (e.g., 2x50 ml falcon tubes) by low-speed centrifugation (~3000 g) for 15 min at room temperature.

6. Carefully pour off the supernatants and gently resuspend in a small amount (~10 ml) of 10% glycerol (prewarmed to room temperature) using disposable 2 ml plastic pipettes. Combine the contents of the two tubes into one and add 10% glycerol to 45 ml. Recentrifuge and wash a second time.

Note: Do not use a vortex mixer to resuspend cells. The vortex will infuse air into the resuspension, resulting in extensive flocculation of the cells. Subsequent centrifugation will not pellet all the cells.
Pellets of cultures pretreated with glycine are often softer than untreated cultures and need to be handled with care.

7. Ensure that the pellet is well-drained. Resuspend the cells fully in 800-1000 µl of 10% glycerol. This is sufficient for four electroporations (200 µl per cuvette).

Note: It is important that the cells are fully resuspended. Clumping of cells significantly reduces their transformation efficiency and increases the risk of arcing.

Electroporation

8. For each electroporation, gently mix 200 µl of resuspended cells with ~1µg DNA (1-5 µl in sterile deionized distilled water or dilute solutions of Tris (5 mM) EDTA (0.5 mM)). Transfer the mixture to a 0.2 cm electrode-gap electroporation cuvette and place in a portable waterbath or heating block at 35-45°C. Allow the cuvette and contents to equilibrate to temperature for several minutes.

Note: In contrast to fast-growing mycobacteria, transformation efficiencies of slow-growing mycobacteria are **substantially improved** by electroporation at elevated temperatures. Electroporation at 37°C can result in efficiencies up to 100-fold higher than those achieved at 0°C (Wards and Collins 1996).

9. Dry the cuvette thoroughly and place in the electroporation chamber inside a biosafety cabinet. Subject to a pulse of 2.5 kV, 25 μF, at a resistance of 1000 Ω.

Note: To avoid potential contamination of the electroporation apparatus, which is kept outside the biosafety cabinet, activate the pulse using knuckles instead of fingertips.

10. **Immediately** after the pulse, add 1 ml of 7H9 broth at room temperature.

Note: This step is important. Immediate dilution of the cells with outgrowth medium allows better recovery from the pulse and results in higher transformation efficiencies.

11. Gently resuspend the sample using a pasteur pipette, transfer to a sterile microfuge tube and incubate at 37°C for ~24 h. This allows establishment of plasmid replication and expression of antibiotic resistance.

12. Pellet the cells in a microfuge for 2-3 min at 6000 g inside a biosafety cabinet. Pour off the supernatant and resuspend by gentle pipetting in 400 μl of 7H9 broth. Ensure that the cells are fully resuspended. If high numbers of transformants are expected, prepare dilutions in 7H9 broth as appropriate.

Cell plating

Note: It is important to keep the intensity and length of pelleting to a minimum. Long centrifugation will result in cell pellets that are difficult to resuspend.
The cells must be thoroughly resuspended to ensure that resistant colonies have arisen from single cells.

13. Spread the cells in 100 μl aliquots onto 7H11 medium containing the appropriate antibiotic selection. Since the transformants may take up to 4 weeks to form colonies, seal the

plates with parafilm or place them in airtight bags to prevent dehydration. Incubate at 37°C.

Results

The figures in Table I illustrate the influence of temperature on the transformation efficiencytemperature of the non-integrating cosmid vector pYUB18 into a range of slow-growing mycobacterial strains. Substantial gains in efficiency can be seen by performing the electroporation at 37°C instead of 0°C. The number of transformants obtained is determined by calculating the number of colonies on the plates spread with cells electroporated with DNA minus the number of colonies on the plates spread with cells alone. The number of spontaneous mutants antibiotic-resistant colonies varies with the vector, the host strain, and the antibiotic used. Spontaneous Kanamycin-resistant colonies of *M.bovis* frequently arise and the number can vary as much as 5-fold between different strains. A high background will make it more difficult to find the desired genetic construct when screening transformants, so the choice of strain is important.

Table 1. Electroporation of slow-growing mycobacteria with pYUB18 at different temperatures

Strain	Efficiency[a]	
	0°C	37°C
M. bovis BCG	3.8×10^3	1.2×10^5
M. tuberculosis H37Rv	8.5×10^3	2.7×10^5
M. intracellulare 96/613	4	2.2×10^3
M. paratuberculosis 989	4	8.3×10^2

[a] Number of transformants per µg DNA per 10^9 cells electroporated.

Troubleshooting

Arcing

Arcing is an additional biosafety hazard when electroporating pathogenic mycobacteria and can result in the lid being blown off the cuvette and aerosolising of the sample inside the electroporation chamber. Extra precautions must be taken including delivering the pulse with the electroporation chamber inside the biosafety cabinet. Factors that increase the likelihood of arcing include:

- Salts in the DNA solution - ensure that the DNA solution contains minimal levels of electrolytes. DNA should be ethanol precipitated and washed with 70% ethanol before being resuspended in sterile deionized distilled water or dilute solutions of Tris-EDTA.

- Moisture on the electrodes - ensure that the electrodes are dried with a clean tissue immediately prior to each pulse.

- High cell density - it is more difficult to remove excess salt and media residues from solutions of high cell density. An additional washing step may be needed.

- Excess cell incubation in the final resuspension solution - the longer that cell suspensions are left prior to electroporation, the greater will be the levels of electrolyte through cell lysis. Cells that have been pretreated with glycine are more prone to this. Perform the electroporation as soon as possible after resuspending the cells.

It is advisable to electroporate the cell control first. If it arcs then the samples containing the DNA will arc as well. A time constant below 12 msec for the cell control indicates arcing is likely to occur with samples containing DNA as these will generally have a lower time constant than cells alone.

The addition of more 10% glycerol (~ 200 µl) to the cuvette can sometimes overcome arcing problems. If arcing results in displacement of the cuvette lid the electroporation chamber and the laminar flow cabinet should be decontaminated using appropriate procedures before further use. It is **strongly advised** to discontinue electroporation experiments using highly virulent strains if explosive arcing occurs.

Comments

The major factor in obtaining high transformation efficiencies with slow-growing mycobacteria is the use of a high temperature for electroporation. In contrast, the efficiency of transformation into *Mycobacterium smegmatis*, a rapid-growing species, and many gram-negative organisms, is higher at 0°C and decreases with temperature (Wards and Collins 1996).

Transformation efficiency varies little with the age of the culture. Cultures of *M.bovis* electroporated with pYUB18 have been shown to yield $1x10^5$ and $1.6x10^5$ transformants per µg DNA per 10^9 cells electroporated when harvested at an OD_{600} of 0.296 and 1.606 respectively (Wards and Collins 1996). Maximum numbers of transformants can be obtained by electroporating more cells.

Acknowledgements. This research was supported by the Foundation for Research, Science and Technology, New Zealand.

References

Wards BJ, Collins DM (1996) Electroporation at elevated temperatures substantially improves transformation efficiency of slow-growing mycobacteria. FEMS Microbiol Lett 145:101-105

Electrotransformation of *Photobacterium damselae* subsp. *piscicida*

JUAN M. CUTRÍN, JUAN L. BARJA and ALICIA E. TORANZO

Introduction

Photobacterium damselae subsp. *piscicida* (formerly *Pasteurella piscicida*) is a marine pathogenic bacterium associated with severe epizootics in cultivated and wild fish. Attempts to transform it using standard electroporation protocols (Dover et al., 1988; Cutrín et al., 1994) were not successful. In these protocols, electroporation efficiency (number of transformants per μg DNA) is normally the parameter to optimize, using high bacterial levels and low nucleic acid concentrations. With this bacterium, efficiency was limited by field strength, so when low field strengths and short pulse lengths were used, transformation efficiencies were below the detection limit, while high field strengths and long pulse lengths resulted in arcing. This problem is common in other marine or estuarine bacteria such as *Caulobacter* sp (Gilchrist and Smit, 1991), *Vibrio* spp (Smigielski et al., 1990), *Vibrio vulnificus* (McDougald et al., 1994) and *Vibrio anguillarum* (Cutrín et al., 1995). When cells were grown in standard saltwater medium and prepared for electroporation with any transformation buffer of low ionic strength, excessive cell lysis oc-

✉ Juan M. Cutrín, Universidad de Santiago de Compostela, Departamento de Mirobiología y Parasitología e Instituto de Acuicultura, Facultad de Biología, Santiago de Compostela, 15706, Spain
(*phone* +34-981-563100(#16052);
fax +34-981-596904; *e-mail* iacuvir@usc.es)
Juan L. Barja, Universidad de Santiago de Compostela, Departamento de Mirobiología y Parasitología e Instituto de Acuicultura, Facultad de Biología, Santiago de Compostela, 15706, Spain
Alicia E. Toranzo, Universidad de Santiago de Compostela, Departamento de Mirobiología y Parasitología e Instituto de Acuicultura, Facultad de Biología, Santiago de Compostela, 15706, Spain

curred. On the other hand, when the cell preparation protocol was isotonic for the bacteria, electroporation was not successful because of arcing.

Because of the difficulty in obtaining high numbers of competent cells, our strategy is to give priority to the frequency of transformation (proportion of viable cells after electroporation that can be transformed by the DNA sample) which is optimized using high DNA concentrations with the purpose of transforming any competent cell that survived the electrical pulse. Therefore, we developed a modified protocol in which the buffers for preparation of competent cells are as isotonic as possible avoiding conductance problems.

Materials

Equipment
- Gene Pulser^{TM} Apparatus (Bio-Rad)
- Pulse controller (Bio-Rad)
- Gene Pulser cuvettes 0.2 cm electrode gap (Bio-Rad)
- Spectrometer Lambda 2 (Perkin Elmer)
- Centrifuge Centrikon T-324 (Kontron)
- Centrifuge bottles 500, 250 and 50 ml (Kontron)
- Microfuge Eppendorf 5414 (Eppendorf)
- Microfuge tubes (Eppendorf)

Plasmids
- Plasmid pCML (Cutrín et al., 1995)

Media
- BHI-1 medium
  - 52 g/l brain heart infusion (Difco)
  - 0.5% NaCl
  - 0.4% glucose
- BHA-1 medium
  - 37 g/l brain heart agar (Difco)
  - 0.5% NaCl
- Nutrient Morita broth medium
  - 0.5% bacto tryptone
  - 0.5% bacto yeast extract
  - 2.5% NaCl
  - 30 mM MgCl$_2$ · 6 H$_2$O
  - 30 mM MgSO$_4$ · 7 H$_2$O
  - 15 mM CaCl$_2$ · 2 H$_2$O

- 1 mM Na_2HPO_4
- 20 mM glucose

- Solution I
  - 2 mM HEPES (pH 6.8)
  - 100 mM sucrose
- Solution II
  - 2 mM HEPES (pH 6.8)
  - 100 mM sucrose
  - 5 mM $CaCl_2$
- Solution III
  - 10% (v/v) glycerol in Milli-Q water

Electropora-
tion buffers

Procedure

1. Using a sterile platinum wire, streak *Ph. damselae* subsp. *piscicida* strain directly from a frozen stock (stored at -80°C in freezing medium) into the surface of an BHA-1 agar plate. Incubate the plate for 24 hours at 25°C.

Preparation of
competent
cells

2. Transfer two or three isolated colonies into 5 ml of BHI-1 broth in a 50 ml flask. Incubate the flask for 16 hours at 25°C.

3. Dilute the culture 1:50 with 500 ml of fresh BHI-1 broth in a 1 liter flask. Grow the bacteria at 25°C for 2-4 hours with gentlly shaking. For efficient transformation it is essential that the cells should be in exponential phase. To monitor the growth of the culture, determine the A_{600} every 30 minutes until a value ranging from 0.5 to 0.8 is achieved.

Note: All subsequent steps in the procedure should be carried out aseptically and on ice. Tubes, rotors and pipettes should be used chilled.

4. Transfer the cells to sterile ice-cold 500 ml bottles (Kontron). Cool the cultures by storing the tubes on ice.

5. Harvest the cells by centrifugation at 4,000 x g for 15 min at 4°C in a Kontron A6.9 rotor (or its equivalent).

6. Discard carefully the media from the cell pellet. Stand the bottles in an inverted position for 1 minute to allow the last traces of media to drain away.

7. Resuspend and wash the pellet in 100 ml of solution I by gently vortexing. Transfer the cells to sterile ice-cold 250 ml tubes (Kontron). Cool the cultures by storing the tubes on ice.

8. Harvest the cells by centrifugation at 4,000 x g for 15 min at 4°C in a Kontron A6.14 rotor (or its equivalent).

9. Resuspend and wash the pellet in 10 ml of solution II by gentle vortexing. Transfer the cells to sterile ice-cold 50 ml tubes (Kontron). Cool the cultures by storing the tubes on ice.

10. Harvest the cells by centrifugation at 4,000 x g for 15 min at 4°C in a Kontron A8.24 rotor (or its equivalent).

11. Working quickly, resuspend and wash the pellet in 1 ml of solution III by gently vortexing. Transfer the cells to sterile ice-cold 1.5 ml microfuge tubes (Eppendorf). Cool the cultures by storing the tubes on ice.

12. Harvest the cells by centrifugation at 12,000 x g for 15 min at 4°C in a centrifuge Eppendorf 5414 (or its equivalent).

13. Immediatelly resuspend the pellet in 100 µl of solution III.

Note: Cell concentration can be calculated by plating on BHA-1 agar medium and it should be arround 10^8-10^{10} cfu/ml.

Transformation protocol

1. Using a chilled, sterile pipette tip, tranfer 50 µl of competent cells (5×10^{10} cfu/ml) to a chilled, sterile microfuge tube.

Note: All pipette tips, pasteur pipettes, microfuges tubes, electroporation cuvettes and buffers needed after this, should be chilled.

2. Add 1 µl of plasmid DNA (1 µg/µl for best results) to the same tube.

Note: DNA plasmids were extracted by alkaline lysis (Birnboim and Doly 1979) of *Escherichia coli* MC1060 cells and purified and purified by equilibrium centrifugation in CsCl-Ethidium bromide gradient (Sambrook et al., 1989). Ethidium bromide was removed from the collected plasmid DNA by extraction with 1-butanol, as well as CsCl by dialysis for 24-48 h against several changes of TE (pH 8.0). Plasmid DNA were precipitated with cold ethanol and resupended in Milli-Q water. The DNA concentration of these stocks was estimated by absorbance at 260nm (assuming that one A_{260} unit is 50 µg/ml) and also by comparing the band intensity on an ethidium bromide-stained agarose gel with known concentrations of λ DNA.

3. Mix the suspension by pipeting. Leave on ice for 1 minute.

Note: Two controls should be included on the experiments; positive control may be competent bacteria which are electrotransformed with a known amount of standard preparation of DNA, and a negative control of competent bacteria without DNA.

4. Set Gene Pulser at 2.5 kV and 25 µF. Set the pulse controller at 200 Ω.

5. Transfer mixture to cold 0.2 cm electroporation cuvette and shake suspension to the bottom of the cuvette.

6. Put the cuvette in the chamber slide and apply a single electric pulse. Check voltage and time constant after pulse.

Note: τ should be around 4.8.

7. Immediately after discharge, remove the cuvette from the chamber slide and add 1 ml of Nutrient Morita broth medium. Resuspend quickly with a pasteur pipette.

8. Tranfer the cells to sterile test tube and incubate with gentle shaking at 25°C for 3 hours.

9. Tranfer 100 µl of transformed competent cells onto agar selective medium: BHA-1 supplemented with 20 mM glucose, and the appropriate antibiotic concentration that inhibit the rest of the cells (chloramphenicol, ampicillin, kanamycin, tetracycline depending on the plasmid used) to identify the transformants.

10. Using a sterile Drigralsky rod, spread the transformed cells over the surface of the agar plate. Leave the plates at room temperature until the liquid has been absorbed.

11. Invert the plates and incubate at 25°C. Colonies should appear in 24 hours.

Results

The effect of electroporation buffers used to prepare competent cells are the most important variables on electroporation of *Ph. damselae* subsp. *piscicida*. All solutions were developed to minimize the conductivity of the suspension and avoid arcing in the sample chamber (Trevors et al. 1991). The presence of glucose in the growth medium used to prepare the competent cells was also important; electroporation efficiency increased approximately 10- fold.

Once electric pulse was applied, the survival rate was always lower than 10%, and the frequency of transformation was approximately 10^{-6}-10^{-7}, values similar to that found in other bacterial species. Plasmid DNA with molecular sizes between 2.6 and 13.7 Kb can be introduced onto several *Ph. damselae* subsp. *piscicida* strains by this protocol. The maximum efficiency achieved was 9.8×10^2 transformants/µg of plasmid DNA (Cutrín et al. 1995).

Although in our experiments, electrotransformation produced only low numbers of transformants, the electroporation protocol constitutes a useful procedure for the genetic manipulation of *Ph. damselae* subsp. *piscicida*.

Troubleshooting

- Absence of transformant colonies on selective medium:
  Low number of competent cells after preparation protocol. Check viable cells before pulse; it should be more than 10^8 cfu/ml.

- Arcing when an electric pulse was applied:
  High ionic strength on the mixture. Be sure that your electroporation solutions are well prepared.

References

1. Birnboim A, Doly J (1979) A rapid extraction procedure for screening recombinant plasmid DNA. Nucleic Acids Res. 7: 1513-1525.
2. Cutrín J L, Conchas R F, Barja J L, Toranzo A E (1994) Electrotransformation of *Yersinia ruckeri* by plasmid DNA. Microbiologia SEM 10: 69-82.
3. Cutrín J L, Barja J L, Toranzo A E (1995) Genetic transformation of *Vibrio anguillarum* and *Pasteurella piscicida* by electroporation. FEMS Microbil. Lett. 128: 75-80.
4. Dower W J, Miller J F, Ragsdale C W (1988) High efficiency transformation of *Escherichia coli* by high voltage electroporation. Nucleic Acids Res. 16: 6127-6145
5. Gilchrist A, Smit J (1991) Transformation of freshwater and marine caulobacters by electroporation. J. Bacteriol. 173: 921-925.
6. McDougald D, Simpson L M, Oliver J D, Hudson M C (1994) Transformation of *Vibrio vulnificus* by electroporation. Current Microbiol. 28: 289-291.
7. Sambrook J, Fristsch E F, Maniatis T (1989) Molecular cloning. A laboratory manual. Cold Spring Harbor Laboratory, Cold Spring Harbor, New York.
8. Smigielski A J, Wallace B, Marshall K C (1990) Genes responsible for size reduction of marine vibrios during starvation are located on the chromosome. Appl. Environ. Microbiol. 56: 1645-1648.
9. Trevors J T, Chassy B M, Dower W J, Blaschek H P (1991) Electrotransformation of bacteria by plasmid DNA. In: Chang D C, Chassy B M, Saunders J A, Sowers A E (eds) Handbook of electroporation and electrofusion. Academic Press, New York, pp 265-289.

Actinobacillus actinomycetemcomitans: Electrotransformation of a Periodontopathogen

KEITH P. MINTZ and PAULA FIVES-TAYLOR

Introduction

Periodontal diseases are chronic inflammatory conditions of the subgingival epithelium that have bacterial etiologies. *Actinobacillus actinomycetemcomitans*, a gram-negative, capnophilic coccobacillus, has been implicated in the pathogenesis of juvenile and adult periodontitis. This bacterium elaborates a number of virulence factors that may contribute to the disease process. Molecular analysis is crucial to understanding the precise role of these and other bacterial factors.

Efficient transformation systems are critical for these analyses. Few transformation systems are available for molecular analysis of *A. actinomycetemcomitans*. Although conjugation is an efficient method for gene transfer, this approach requires the development of antibiotic resistant recipient cells, specific donor cells and specific host and plasmids that contain mobility elements. Electrotransformation eliminates all of the above concerns. In this protocol we will discuss an efficient electroporation system that has been developed for the transformation of *A. actinomycetemcomitans*.

Keith P. Mintz, University of Vermont, Department of Microbiology and Molecular Genetics, The Markey Center for Molecular Genetics, College of Medicine and College of Agriculture and Life Sciences, Burlington, Vermont, 05405, USA

✉ Paula Fives-Taylor, University of Vermont, Department of Microbiology and Molecular Genetics, The Markey Center for Molecular Genetics, College of Medicine and College of Agriculture and Life Sciences, Burlington, Vermont, 05405, USA (*phone* +01- 802-656-1121; *fax* +01-802-656-8749; *e-mail* pfivesta@zoo.uvm.edu)

Materials

- Gene Pulser Apparatus and Pulse Controller (Bio-Rad La-   **Equipment**
  boratories) or similar device.
- CO_2 Incubator or Anaerobic Jar System with carbon dioxide
  generator gas pack (VWR Scientific Products)
- 2 mm gap electroporation cuvette (BTX)

- Trypticase Soy Broth containing Yeast Extract (TSBYE)   **Media and**
  - 30 g/liter Trypticase soy broth   **buffers**
  - 6 gm/liter yeast extract
  - Added to one liter of distilled water
  - Autoclave
- Electroporation buffer
  - 272 mM sucrose
  - 2.43 mM K_2HPO_4
  - 0.57 mM KH_2PO_4
  - 15% glycerol
  - Filter sterilize and store at 4°C.

Procedure

Making frozen competent cells

Bacterial cell density and phase of growth is very important in making highly competent cells. Therefore, to achieve the best results, the bacteria are cultured in a step-up fashion.

1. Inoculate 0.1 ml of frozen stock to 1 ml TSBYE supplemented with 50 µg/ml bacitracin and 5 µg/ml vancomycin.

2. Incubate at 37°C in an incubator with an atmosphere containing 10% CO_2 (standard incubation) or Gas Pak system overnight.

3. Transfer 1 ml of the overnight culture to 20 ml of TSBYE supplemented with 50 µg/ml bacitracin and 5 µg/ml vancomycin. Incubate in standard conditions.

4. Add 20 ml of fresh medium to the overnight culture and incubate in standard conditions for 3 hours or to an optical density of 0.3-0.4 as measured at 600 nm. The cells should

be in early- to mid-logarithmic phase of growth. The growth phase of the bacteria is a very important parameter in the transformation of *A. actinomycetemcomitans*.

5. Determine the optical density (600 nm) of the culture and record (see below).

6. Centrifuge the bacteria at 4000 x g for 15 minutes at 4°C.

Note: All of the following steps should be carried out at 4°C using prechilled pipettes and reagents.

7. Resuspend the bacterial pellet in 20 ml of electroporation buffer and centrifuge as stated above in step #6.

8. Decant the supernatant and repeat step #7.

9. Resuspend the bacterial pellet to an optical density of 6.0.

Note: To determine this value: first, divide 6.0 by the optical density recorded in step #5. This is the concentration factor. Second, divide the original volume of the starting culture by the concentration factor. This value is the volume of electroporation buffer used to resuspend the final bacterial pellet.

10. Aliquot 250 µl of the cell suspension into vials and snap freeze the vials in a methanol/dry ice bath. Store at -70°C.

Electroporation of competent *A. actinomycetemcomitans* cells

Cuvettes should be stored at -20°C before use. Machine settings: 2.5 kV, 200 ohms and 25 µFarads.

1. Thaw competent cells on ice.

2. Distribute 50 µl of cells into ice-cold microcentrifuge tubes.

3. Add DNA to cells. The volume of the DNA solution added to the cuvette will depend on the salt concentration of the DNA solution. High concentrations of salt will cause arcing.

4. Transfer the DNA/cell mixture to an ice-cold electroporation cuvette. Remove all air bubbles from the electroporation cuvette by tapping.

5. Place the cuvette in the cuvette carrier and pulse following the manufacturer's instructions. Record the time constant.

6. Immediately after pulsing, add 1 ml of pre-warmed TSBYE to the cuvette.

7. Transfer the contents of the cuvette to a culture tube and allow the cells to recover for 3 hours in standard conditions.

8. Plate varying amounts of the cell suspension on TSBYE agar plates containing the appropriate antibiotic. Incubate in standard conditions for at least 48 hours.

Results

Electrotransformation is a rapid and efficient method for the transfer of DNA into bacteria. The protocol described above has been optimized to obtain the highest transformation efficiencies. An important parameter to be considered for effective transformation of A. *actinomycetemcomitans* is the stage of

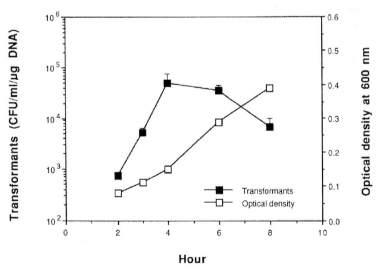

Hour

Fig. 1. Effect of the stage of bacterial growth on transformation. Bacteria were grown to different stages in their growth cycle before electroporation. Closed symbols represent the number of transformants per milliliter per microgram of plasmid DNA. Open symbols indicate the OD at 600 nm of the bacteria culture at the time of harvesting.

growth of the bacteria used for electroporation. Transformation was highly dependent on the stage of bacterial growth (Figure 1). The highest transformation efficiencies were achieved when the bacteria were in the early logarithmic phase of growth. Bacteria harvested in the stationary phase of the growth cycle remained transformable, although at much lower efficiencies.

The concentration of DNA used in transformation is also critical (Figure 2). Addition of increasing concentrations of plasmid DNA (ranging from 0.1 to 10 µg/ml) resulted in increasing number of transformants. The maximum number of transformants were obtained at 2.5 µg/ml plasmid DNA. The transformation efficiency remained constant with DNA concentrations ranging from 0.5 to 3.5 µg/ml.

Under optimal conditions, the efficiency of transformation of *A. actinomycetemcomitans* using a replicating plasmid is between 10^4 to 10^5 transformants per µg of DNA. This level of transformation is comparable to other oral bacteria. An excellent pa-

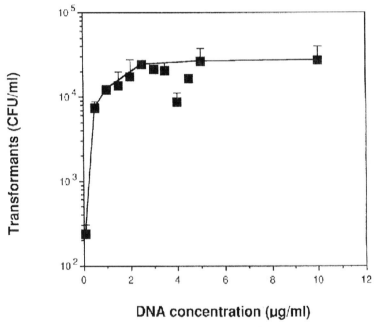

Fig. 2. Effect of plasmid DNA concentration on transformation. DNA concentrations of 0.1 to 10 µg/ml were added to the bacterial samples. Concentrations of DNA were added in a constant volume of 1 µl. Symbols indicate the number of transformants per milliliter recovered at each DNA concentration.

per that describes in detail the parameters of the transformation of *A. actinomycetemcomitans* can be found in Sreenivasan et al. (1991).

Acknowledgements. This work was supported, in part, by Public Health Service grant R01DE09760.

References

Sreenivasan PK, LeBlanc DJ, Lee LN and Fives-Taylor PM (1991) Transformation of *Actinobacillus actinomycetemcomitans* by electroporation, utilizing constructed shuttle plasmids. Infect Immun 59:4621-4627

Suppliers

Bio-Rad Laboratories
2000 Alfred Nobel Drive
Hercules, CA 94547 USA
phone: 800-424-6723
fax: 1-800-879-2289

VWR Scientific Products (International)
P.O. Box 1002
Plainfield, NJ 07080 USA
phone: 908-757-4045
fax: 908-757-0313

BTX
11199 Sorrento Valley Road
San Diego, CA 92121-1334 USA
phone: 800-289-2465
fax: 619-597-9594

Francisella in Medical and Veterinary Applications

Siobhán C. Cowley and Francis E. Nano

Introduction

Francisella tularensis, the causative agent of the zoonotic disease tularemia, is an encapsulated Gram negative cocco-bacillus that is a facultative intracellular pathogen. Tularemia is caused by at least two biotypes of *F. tularensis*, designated type A (*F. tularensis tularensis*) and type B (*F. tularensis palaeartica*). *F. tularensis* type A is limited to North America and causes a more severe form of disease. Conversely, the less virulent *F. tularensis* type B is found in North America, Europe, and Asia. Both the type A and type B strains of *F. tularensis* are highly infectious and can initiate infections with an inoculum of less than 10 organisms.

An *F. tularensis* infection is capable of causing a wide variety of symptoms which vary as a function of the route of infection and the immune response of the host. Human infection is initiated by a variety of routes, including the bite of an arthropod vector, direct contact with small wild game (especially rabbits), ingestion of contaminated water or meat, or inhalation of contaminated dust. Infection initiated by a cut or insect bite results in a form of tularemia called ulcero-glandular, which is characterized by the development of an ulcer at the site of infection, followed by a fever and swelling of local lymph nodes. This may progress to other forms of the disease depending on the natural immunity of the host. Indeed, the development of septicemia may lead to infection of the lungs (pneumonic tularemia)

Siobhán C. Cowley, University of Victoria, Department of Biochemistry and Microbiology, Victoria, B.C., V8W 3P6, Canada

✉ Francis E. Nano, University of Victoria, Department of Biochemistry and Microbiology, Victoria, B.C., V8W 3P6, Canada (*phone* 250-721-7074; *fax* 250-721-8855; *e-mail* fnano@uvic.ca)

and a systemic form of the disease, termed typhoidal tularemia. The disseminated typhoidal form of tularemia may also develop as a result of inhalation or ingestion of *F. tularensis*, and a primary pneumonic form may be initiated through aerosol exposure.

DNA homology and 16S RNA phylogenetic studies have shown that *F. tularensis* biotype *novicida* is closely related to other *F. tularensis* biotypes (types A and B) (Hollis et al. 1989). Although *F. novicida* is weakly pathogenic in humans it causes an experimental infection in mice which resembles that of human tularemia. *F. tularensis* strains have been shown to grow within macrophages during experimental murine infections within an unfused acidified phagosome (Anthony et al. 1991a; Fortier et al. 1995). Clearance of murine tularemia can be effected by either CD4$^+$ or CD8$^+$ T cells, and survival is dependent on production of the cytokines interferon-γ, TNF-α, and IL-12 (Yee et al. 1996; Conlan et al. 1994; Anthony et al. 1989; Leiby et al. 1992; Elkins et al. 1996; Elkins et al. 1993). Intramacrophage growth of *Francisella* may be limited by both nitric oxide-dependant and -independant killing mechanisms of the macrophage (Anthony et al. 1992; Polsinelli et al. 1994).

Materials

- Growth media and antibiotic resistance markers
  Francisella strains are grown in trypticase soy broth (Difco) containing 0.1% cysteine (TSB-C) for electroporation. Transformants are selected on cysteine heart agar (Difco) supplemented with 5% defibrinated horse blood (CHA-H) and the appropriate antibiotic. The antibiotic resistance markers which have been reported to successfully express in *Francisella* include the Kanamycin (Km) cassette derived from Tn*903* for both the LVS and *novicida* biotypes (Anthony et al. 1991b), the tetracycline and chloramphenicol resistance genes from the *Escherichia coli* plasmid pBR328 for the LVS biotype (Norqvist et al. 1996), and the erythromycin resistance encoded by the *ermC'* gene from Tn*Max2* for *F. novicida*. Agar plates should be supplemented with kanamycin sulfate (5 µg/ml for *F. tularensis* LVS or 15 µg/ml for the *novicida* biotype), tetracycline (10 µg/ml), chloramphenicol

(2.5 µg/ml), or erythromycin (25 µg/ml). Buffer used for washing cells prior to electroporation consists of 500 mM sucrose.

– *Francisella* strains and safety considerations
The most commonly studied *Francisella* strain is *F. tularensis* live vaccine strain (LVS, ATCC 29684) which is safe for use in a Biosafety Level 2 (BL2) containment laboratory. Conversely, the containment requirements for *F. tularensis* type A and type B strains differs among different countries, but we recommend Biosafety Level 3 containment, and individuals working with these strains should be vaccinated for their own protection. *F. tularensis* biotype *novicida* U112 (ATCC 15482) has been proven to be safe to use under BL2 conditions. A capsule-deficient mutant of *F. tularensis* LVS called LVSR may also be used for electroporation. All strains may be stored as bacterial cultures mixed 1:1 with 2.6% sterile gelatin at -76 °C.

– Plasmids and transposons
It has been hypothesized that *Francisella* sequences are required for efficient *Francisella* transformation (Anthony et al. 1991b). Thus, all plasmids discovered to date which result in stable transformation of *Francisella* strains are either derived from a natural *Francisella* plasmid (Norqvist et al. 1996), or are plasmids containing a cloned insert of *Francisella* DNA. It is important to note that electroporation efficiency of recombinant *Francisella* clones can vary by as much as 10,000-fold with different *Francisella* genetic loci. Furthermore, the location of the selectable marker in the *Francisella* locus may also affect transformation. Therefore it is suggested that established high efficiency transforming *Francisella* clones be used as a positive control. Plasmids pFEN504-3 and pLA68-11 are recommended for the LVS and *novicida* biotypes, respectively (Anthony et al. 1991b). Plasmid pFEN504-3 (ATCC 40849) is a pUC18-derived recombinant plasmid that has a 9 kb LVS insert, with a kanamycin resistance cassette located within the insert DNA. Plasmid pLA68-11 consists of a 6 kb *F. novicida* DNA insert disrupted by a Km cassette cloned into pUC19.

Procedure

1. Grow *Francisella* overnight in TSB-C. The following morning, this culture should be diluted and grown to 100 Klett units (A $_{600}$ of 0.2) using the number 47 filter (540 nm). Cultures of LVS require at least 10^5 cells/ml in order to start growth quickly.

2. Wash cells three times in 500 mM sucrose. After the final wash the cells should be concentrated approximately 300-fold in 500 mM sucrose. An example of a protocol for the electroporation of four samples is as follows: centrifuge 40 ml of culture at 10,000 x g for 5 minutes and resuspend the pellet in 10 ml of sucrose. Wash the cells again and concentrate the cells to 1 ml in sucrose. Move the cell suspension into a screw-cap microfuge tube. Dip the microfuge tube in disinfectant and centrifuge for 5 minutes. Resuspend the pellet in 120 µl of sucrose by pipetting up and down repeatedly. The cell suspension should be kept on ice until needed for electroporation.

3. Add 1 µl (approximately 1 µg) of purified plasmid DNA to 40 µl of washed cells and transfer to a 0.1 cm gap electroporation cuvette. Electroporate at 15 kV / cm at 400 Ohms, with the capacitance set at 25 µF. If HDAP-treated *F. tularensis* are to be used (see Troubleshooting section), then a lower electric field strength of 7.5 kV / cm is recommended. Within 90 seconds after electroporation, add 1 ml TSB-C and grow cells at 37 °C with shaking for 6 hours (for LVS) or 2 hours (for biotype *novicida*).

4. Plate cells on CHA-H plates supplemented with the appropriate antibiotic. *F. novicida* transformants typically appear within 1-2 days and LVS transformants should appear within 3-4 days.

Results

Typical electroporation results for the positive control plasmids pLA68-11 and pFEN504-3 consist of a transformation frequency of 2 x 10^3 transformants per µg DNA for *F. tularensis* LVS, 3 x 10^3

transformants per μg DNA for HDAP-treated *F. tularensis* LVS, and 6 x 10³ transformants per μg DNA for *F. tularensis* LVSR. Electroporation of *F. novicida* is highly inefficient and results only in rare transformants.

Troubleshooting

Capsule negative strains of *F. tularensis* LVS have been shown to electroporate at higher efficiencies and more consistently (Anthony et al. 1991b). A capsule negative variant of LVS (called LVSR) is available (Sandström et al. 1988), or it is also possible to physically remove the capsule from *F. tularensis* LVS with a mild detergent treatment. Electroporation efficiency may be increased by vortexing cells in sucrose containing 1% N-hexadecyl-N,N-dimethyl-3-ammonio-1- propanesulfonate (HDAP) for 30 seconds during the first wash step.

Comments

- The events involved in transformation of *Francisella* are poorly understood. *F. novicida* transforms approximately 1,000-fold better by chemical transformation than by electroporation, whereas *F. tularensis* transforms 1,000-fold better by electroporation (Anthony et al. 1991b).

- Although the *novicida* and LVS biotypes do not usually cause disease in healthy humans, precautions should be taken to avoid the generation of aerosols. Manipulations such as vortexing, opening of centrifuge tubes, and the electroporation itself which may result in arcing should be performed in a type 2 biosafety cabinet. To avoid arcing, DNA should previously have been desalted by drop dialysis or washing with 70% ethanol.

- Growth of *Francisella* cultures in TSB-C yields cells that are transformed better than cultures grown in Chamberlain's defined medium. This phenomenon is attributed to the presence of glucose in Chamberlain's media which increases the production of capsule (Cherwonogrodzky et al. 1994).

- Insufficiently rinsed glassware may contain detergents which can apparently inhibit LVS growth. *Francisella* cultures stored on agar plates survive for less than two weeks at room temperature.

- *F. tularensis* LVS DNA is not methylated and there does not appear to be a DNA restriction system in either *F. tularensis* LVS or biotype *novicida*.

References

Anthony LSD, Burke RD, Nano FE (1991a) Growth of *Francisella* spp. in rodent macrophages. Infect Immun 59: 3291-3296.

Anthony LSD, Ghadirian E, Nestel FP, Kongshavn PAL (1989) Experimental murine tularemia caused by *Francisella tularensis*, live vaccine strain: a model of acquired cellular existence. Microb Path 2: 3-14.

Anthony LSD, Gu M, Cowley SC, Leung WWS, Nano FE (1991b) Transformation and allelic replacement in *Francisella* spp. J Gen Microbiol 137: 2697-2703.

Anthony LSD, Morrissey PJ, Nano FE (1992) Growth inhibition of *Francisella tularensis* live vaccine strain by IFN-gamma-activated macrophages is mediated by reactive nitrogen intermediates derived from L-arginine metabolism. J Immunol 148: 1829-1834.

Cherwonogrodzky JW, Knodel MH, Spence MR (1994) Increased encapsulation and virulence of *Francisella tularensis* live vaccine strain (LVS) by subculturing on synthetic medium. Vaccine 12: 733-775.

Conlan JW, Sjöstedt A, North RJ (1994) CD4[+] and CD8[+] T-cell-dependant and -independant host mechanisms can operate to control and resolve primary and secondary *Francisella tularensis* LVS infection in mice. Infect Immun 62: 5603-5607.

Elkins KL, Rhinehart-Jones T, Culkin SJ, Yee D, Winegar RK (1996) Minimal requirements for murine resistance to infection with *Francisella tularensis* LVS. Infect Immun 64: 3288-3293.

Elkins KL, Rhinehart-Jones T, Nacy CA, Winegar RK, Fortier AH (1993) T-cell-independant resistance to infection and generation of immunity to *Francisella tularensis*. Infect Immun 61: 823-829.

Fortier AH, Leiby DA, Nacy CA, Meltzer MS (1995) Growth of *Francisella tularensis* LVS in macrophages: the acidic intracellular environment provides essential iron required for growth. Infect Immun 63: 1428-1483.

Hollis DG, Weaver RE, Steigerwalt AG, Wenger JD, Moss CW, Brenner DJ (1989) *Francisella philimiragia* comb. nov. (formerly *Yersinia philimiragia*) and *Francisella tularensis* biogroup *novicida* (formerly *Francisella novicida*) associated with human disease. J Clin Microbiol 27: 1601-1608.

Leiby DA, Fortier AH, Crawford RM, Schreiber RD, Nacy CA (1992) In vivo modulation of the murine immune response to *Francisella tularensis* LVS by administration of anticytokine antibodies. Infect Immun 60: 84-89.

Norqvist A, Kuoppa K, Sandström G (1996) Construction of a shuttle vector for use in *Francisella tularensis*. FEMS Immunol Med Microbiol 13: 257-160.

Polsinelli T, Meltzer MS, Fortier AH (1994) Nitric oxide-independant killing of *Francisella tularensis* by IFN-γ-stimulated alveolar macrophages. J Immunol 153: 1238-1245.

Sandström G, Löfgren S, Tarnvik A (1988) A capsule-deficient mutant of *Francisella tularensis* LVS exhibits enhanced sensitivity to serum but diminished sensitivity to killing by polymorphonuclear leukocytes. Infect Immun 56: 1194-1202.

Yee D, Rhinehart-Jones TR, Elkins KL (1996) Loss of either CD4[+] or CD8[+] T cells does not affect the magnitude of protective immunity to an intracellular pathogen, *Francisella tularensis* strain LVS. J Immunol 157: 5042-5048.

Suppliers

Difco Laboratories
P.O. Box 331058
Detroit, MI 48232-7058 USA
Phone: 313-462-8500
FAX: 313-462-8517

Electroporation of the Anaerobic Rumen Bacteria *Ruminococcus albus*

PIER SANDRO COCCONCELLI

Introduction

The species *Ruminococcus albus* is one of the bacteria most active in cellulolysis and xylanolysis composing the complex microbial association of the rumen. Since ruminococci have a relevant role in the hydrolyzation of plant polysaccharides in the rumen, they may represent the bacteria of choice for genetic manipulation aimed at enhancing the digestibility of ruminant (Gregg et al. 1996). Different gene transfer systems have been reported for ruminococci, protoplast fusion (Ohmiya, 1990), conjugal transfer of the plasmid pAMβ1 from *Enterococcus faecalis* to *R. albus* (Aminov et al. 1994) and high voltage electrotransformation (Cocconcelli et al., 1992). This latter technique was used to introduce plasmid vectors in *R. albus* (Cocconcelli et al. 1992) and for the cloning of heterologous β,1-4 glucanase in this group of ruminal cocci (Cappa et al. 1997). In different Gram positive bacteria, the recombinant DNA studies have been greatly facilitated by the development of a host-vector system based on the use of origins of replication from native plasmids. Plasmids, with dimensions ranging from 4 to 70 kb, were isolated by several authors from ruminococci (Asmundson and Kelly 1987; Ohmiya et al. 1990, Anapuma et al. 1996). Although some of these plasmids have been associated with the cellulolytic activity, the function of most of them remains cryptic. Since genetic information on extrachromosomal replicons from ruminococci have not been not available, plasmid harbouring broad

Pier Sandro Cocconcelli, Università Cattolica del Sacro Cuore, Istituto di Microbiologia, via Emilia Parmense 84, Piacenza, 29100, Italy (*phone* +39-523-599248;
fax +39-0523-599246; *e-mail* cocconc@pc.unicatt.it)

host range origin of replication, suitable for Gram positive bacteria have been used for gene cloning in *R. albus* (Cocconcelli et al. 1992; Cappa et al.1997). In particular two types of replicons have been shown to be functional in *R. albus*: the plasmids replicating via single stranded DNA intermediate, belonging to the pSL1 family (del Solar et al. 1993), such as pCK17 (Gasson and Anderson 1985), pPSC22 (Cocconcelli et al. 1997) and pGK12 (Kok et al. 1984), and the pIL253 vector (Simon and Chopin) derived from the broad host range conjugative plasmid pAMβ 1, which have previously been demonstrated to be functional in *R. albus* (Aminov et al. 1994).

The purpose of this chapter is to present a protocol for plasmid transformation *R. albus* by means of high voltage electroporation. The key steps of the electroporation procedure of *R. albus* are presented and discussed.

Outline

An outline of the electroporation protocol for plasmid transformation of *R. albus* is shown in the following:

- Inoculate (1%) one tube of M10. Starting from this tube make serial 1/10 dilution in fresh M10 tubes (8 tubes) and incubate the cultures overnight at 39°C.

- Select the last tubes which present a A_{560} close to 0.5-0.7 (strain dependent value).

- Use 2 ml of the selected culture to inoculate 8 ml M10 tube.

- When there is an increase of 100-150%, harvest the cells by centrifugation at 6000 xg for 10 min.

- Wash the cells twice with 5 ml of ice cold 100 mM NaCl.

- Resuspend the cells in 1.5 of ice chilled electroporation buffer and transfer to a 1.5 ml microcentrifuge tube.

- Centrifuge at 8000 rpm for 5 min and wash with the same buffer.

- Resuspend the cell in 400 µl of electroporation buffer, add the DNA and transfer in the electroporation cuvette.

- Electroporate the cells at 12500 V/cm 25 µF and 400 Ω (ohms).

- Immediately dilute with 1 ml of M10bsuc and incubate for 2 hours at 39°C.

- Spread the cells on M10asuc with antibiotics.

Materials

The bacterial strains and the plasmid used for electrotransformation are listed in Table 1. The *R. albus* strains were cultivated in M10 medium (Caldwell and Bryant 1966) in anaerobic conditions using an anaerobic cabinet (atmosphere: 10% CO_2, 5% H_2, 85% N_2). Concentration of antibiotics in selective media for ruminococci were 5 µg/ml and 10 µg/ml for erythromycin and chloramphenicol respectively. Plasmid DNA was extracted and purified from *R. albus* following the protocol described by Anapuma et al.(1996). The single stranded intermediate plasmid vectors pCK17, pPSC22, and pGK12 used for electroporation were extracted from *E. coli* using the purification cartridges of Quiagen GmbH, (Germany) following the supplier's instructions. Plasmid pIL253 was isolated from *Lactococcus lactis* subsp. *lactis* as described by Anderson and McKay (1983).

Bacterial cultures, media, plasmids

Table 1. Bacterial strains and plasmids

Strains /plasmid	Source or reference
R. albus ATCC 2752	ATCC
R. albus RC6	Cocconcelli et al. 1992
R. albus RC16	Cocconcelli et al. 1992
R. albus RC6	Cocconcelli et al. 1992
R. albus RC6	Cocconcelli et al. 1992
R. albus BF18	Cappa et al. 1997
R. albus BF21	Cappa et al. 1997
pCK17	Gasson and Anderson 1985
pPSC22	Cocconcelli et al. 1996
pGK12	Kok et al. 1985
pIL253	Simon and Chopin 1988

Electroporation buffer: 272 mM Sucrose

Media	reference
M10	Caldwell and Bryant (1966)
M10 broth added with 0.25 M sucrose (M10bsuc)	this work
M10 agar added with 0.25 M sucrose (M10asuc)	this work

Equipment   – Gene Pulser^{TM} and Pulse Controller^{TM} (Bio Rad)
        – Anaerobic glove box (Forma Scientific, Marietta, Ohio, USA)

Procedure

All the steps from 1 to 4 are performed in the anaerobic glove box using pre-reduced plasticware and glassware . During the steps taken outside glovebox plastic centrifuge tubes must be used which have previously been tested to see that an anaerobic atmosphere can resist inside. It is suggested to test the tubes using a solution containing resarzurine as indicator of the redox conditions.

1. Using a log phase culture of *R. albus*, inoculate (1%) one tube containing 10 ml of freshly prepared M10. Starting from this culture make a serial decimal dilution in tubes containing 9 ml of pre-warmed M10. In these conditions it is possible to achieve a series of tubes inoculated with a decreasing number of bacterial cells. Incubate the cultures overnight at 39°C.

2. Among the series of cultures, select the tube which presents an optical density (A_{560}) close to 0.6.

3. Inoculate with 2 ml of the selected culture two 8 ml of pre-warmed M10 in a 13 ml polypropylene centrifuge tube and incubate this culture at 39°C.

4. Fill a 2 ml sterile plastic syringe, previously reduced in the anaerobic glove box, with 1 ml of M10bsuc

5. Use one of the two tubes to check the optical density and when the A_{560} increases by 100-150% transfer the second

tube outside the anaerobic cabinet, chill the culture on ice and harvest the cells by centrifugation at 6000 x g for 10 min. After the spinning, transfer the tube into the anaerobic glove box. In most cases it is very difficult to pellet the cells of *R. albus*: for this reason it is suggested to carefully eliminate the supernatant using a sterile 10 ml pipette.

6. Wash the cells with 5 ml of ice cold 100 mM NaCl and harvest by centrifugation at 6000 x g for 10 min, using the pipette to aspirate the supernatant after each centrifugation.

7. Resuspend the cells in 1.5 ml of ice chilled electroporation buffer and transfer to a 1.5 ml microcentrifuge tube. Centrifuge at 8000 rpm for 5 min.

8. Carefully remove the supernatant with a micropipette and repeat step 7.

9. Resuspend the cell in 400 µl of electroporation buffer, add the DNA (200 ng in 2 µl), mix gently and transfer in the 0.2 cm gap electroporation cuvette

10. Close the electroporation cuvette with a sterilized rubber cap, previously cut to the appropriate size, as shown in Figure 1 and transfer outside the cabinet.

11. Electroporate the cells at 12500 V/cm 25 µF and 400 Ω (ohms)

12. Immediately punch the rubber plug with the sterile syringe and inject 1 ml of M10bsuc in the cuvette, as shown in Figure 1. Mix gently and return the cuvette to the anaerobic conditions.

13. Incubate for 2 hours at 39°C to allow the expression of new introduced genes.

14. Spread the cells on M10asuc added with selective agents.

Results

In several reports of electroporation of Gram-positive bacteria it was shown that optimal transformation frequencies were dependent on the growth phase of the cells, on the field strength, on the time constant and on the buffer used. The present study was in-

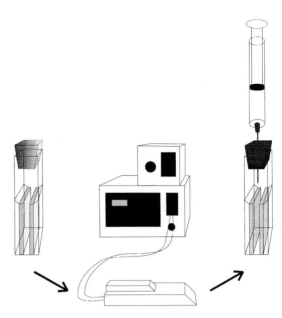

Fig. 1. The cuvettes and the dilution system used for the electroporation in anaerobic condition. A rubber plug is used to seal the cuvette outside the anaerobic glove box. After the electroporation the dilution is achieved using a syringe containing M10 broth added with 272 mM sucrose.

itiated by determining the composition of an electroporation buffer that would produce a higher number of transformants in *R. albus*. The following factors, pH (from 4 to 7), sucrose concentration (0,272 mM and 500mM), polyethylene glycol 6000 (0 and 10%) were tested using an early log phase culture of *R. albus*. The highest frequencies of electrotransformation were achieved using a non-buffered sucrose solution (272 mM). The effect of field strength and time constant was studied using 0.2 cm gap cuvettes. Transformation was more efficient at high field strength: the number of transformants increased progressively for 7500 Vcm^{-1} to12500 Vcm^{-1}, while electroporated cells were not detected using electric fields lower than 7500 Vcm^{-1}. The highest frequencies of electroporation at electric field of 12500 Vcm^{-1} were obtained setting the Bio-Rad Pulse Controller at 400 Ω, so achieving time constant close to 6-7 ms.

An additional factor with a striking effect on the electroporation frequencies was the selection of the correct growth phase, which was dependent on the strain considered. In general all the seven strains tested were more efficiently transformed when early log phase cultures (A_{560} = 0.3) were used instead of late log phase ruminococcal cells (A_{560} = 0.8).

Since each strain tested showed a day-to-day variation in the frequency of transformation, in order to standardize the growth condition of *R. albus* a series of over-night cultures was inoculated with a decreasing number of cells. The culture, among the inoculated tubes, showing an optical density close to 0.6 (from 0.5 to 0.7 depending on the strain) was then used to inoculate (20%) a fresh tube of pre-warmed M10, which was then incubated until the A_{560} increased by 100-150%. For the strains considered by us, this approach yielded a culture in the appropriate growth phase for the electrotransformation. The use of this protocol of cell preparation permitted the reduction of the day-to day variation and gave the best results for all the strains considered.

When cells of *R. albus*, prepared as described in the Procedure were frozen in electroporation buffer at -80°C, the frequency of transformation sharply decreased when compared with cells from the same batch before freezing.

The quantity of DNA is another factor affecting the frequency of electroporation. DNA concentration between 10 and 200 ng/ml showed a concentration dependent increase in electrotransformation frequency. Higher concentration resulted in a saturation, without an increase in the number of electrotransformants.

However, the most critical factor affecting the results of the electrotransformation experiments in *R. albus* is the maintenance of a strictly anaerobic atmosphere during all the steps from the collection of cells to the plating after electroporation. Any contact with oxygen results in a strong inhibition of the *R. albus* cells and in the relevant decrease of the transformation frequency. For this reason care must be taken in pre-reducing all the plasticware and glassware and in all the steps performed outside the anaerobic cabinet.

The final protocol described in the Procedure and in the Outline sections represents the optimal condition for the transformation of *R. albus* among the different conditions tested by us (Cocconcelli et al. 1992, Cappa et al 1997). These conditions allowed us introduce plasmid DNA into strain of *R. albus* with frequencies ranging from 10 to 10^5 transformants/µg of purified plasmid DNA.

References

Anderson D, McKay LL (1983) Simple and rapid method for isolating large plasmid DNA from lactic streptococci. Appl Environ Mcrobiol 46:549-552

Aminov RI, Kaneichi K, Miyagi T, Sakka K, Ohmiya K(1994) Construction of genetically marked *Ruminococcus albus* strains and conjugal transfer of plasmids pAM(1 into them. J Ferment Bioeng 78:1-5

Anupuma V, Grover S, Batish VK (1996) Comparative evaluation of methods for isolation of plasmids from ruminococci of buffalo rumen origin. Microbiologie Aliments Nutrition 14:113-123

Asmundson RV, Kelly WJ. (1987) Isolation and characterization of plasmid DNA from *Ruminococcus* Current Microbiol 16:97-100

Caldwell DR, Bryant MP(1966) Medium without rumen fluid for non selective enumeration and isolation of rumen bacteria. Appl Microbiol 14:794-802

Cappa F, Riboli B, Rossi F, Callegari ML Cocconcelli PS (1997) Construction of novel *Ruminococcus albus* strains with improved cellulase activity by cloning of *Streptomyces rochei* endoglucanase gene. Biotech Lett 19:1151-1155

Cocconcelli PS, Ferrari E, Rossi F, Bottazzi V (1992) Plasmid transformation of *Ruminococcus albus* by means of high-voltage electroporation. FEMS-Microbiol Lett 94:203-207.

Cocconcelli PS, Elli M, Riboli B, Morelli L(1996) Genetic analysis of the replication region of the *Lactobacillus* plasmid vector pPSC22 . Res Microbiol 147: 619-624.

Del Solar G, Moscoso M, Espinosa M (1993) Rolling circle-replicating plasmids from Gram-positive and Gram negaive bacteria: a wall falls. Mol Microbiol 8: 789-796.

Gasson MJ, Anderson PH (1985) High copy number plasmid vectors for use in lactic streptococci. FEMS Microbiol Lett 30: 193-196

Gregg K, Allen G, Beard C (1996) genetic manipulation of rumen bacteria from potential to reality. Aus J Agr Res 47:247-256.

Kok J, van der Vossen JM, Venema G (1984) Construction of plasmid cloning vectors for lactic streptococci which also replicate in *Bacillus subtilis* and *Escherichia coli*. Appl Environ Microbiol, 48:726-731

Ohmiya K (1990 Genetic engineering in rumen anaerobic bacteria. In Oshino S, Onodera R, Minato H, Itabashi H (eds) : The rumen ecosystem: the microbial metabolism and its regulation. Japan Scientific Societies Press. Tokio, pp 195-202.

Ohmiya K, Hoshino C, Shimuzu S (1990) Novel plasmid responsible for the cellulose utilizing ability of *Ruminococcus albus*. Jpn J Zootech Sci 61:557-561

Simon D, Chopin A (1988) I: Construction of a vector plasmid family and its use for molecular cloning in Streptococcus lactis. Biochimie 70:559-566.

Electroporation of *Legionella* Species

V.K. VISWANATHAN and NICHOLAS P. CIANCIOTTO

Introduction

The molecular analysis of *Legionella pneumophila* had been limited by the lack of methods to genetically manipulate the organism. Since bacteriophages and transduction have not been reported in *L. pneumophila*, methods of introducing DNA into the organism were limited to conjugation until 1990. Since then electroporation has rapidly replaced conjugation as the preferred method of introducing DNA into *L. pneumophila* for mutagenesis - mediated by transposons (Pope et al, 1994, Wiater et al, 1994, Pruckler et al, 1995, McClain and Engleberg, 1996, Hickey and Cianciotto, 1997) or by allelic exchange (Cianciotto and Fields, 1992, Sadosky et al, 1993, Abu Kwaik et al, 1997) and for complementation studies (Marra and Shuman, 1992, Berger et al, 1994, Abu Kwaik and Pederson, 1996).

Electroporation is an easier and more efficient alternative to conjugation for the introduction of plasmids into *L. pneumophila*. While high frequency of conjugation could be achieved only with restriction mutants of *L. pneumophila*, electroporation has been successfully used to transform plasmids into wild type *L. pneumophila* at high frequencies. Some of the plasmid types that have been successfully' transformed into *L. pneumophila* are IncP and IncQ derivatives (Marra and Shuman, 1992), plasmids containing *ori*RSF1010 (Berger and Isberg, 1993, Brand et al,

V.K. Viswanathan, Northwestern University, Department of Microbiology and Immunology, 303 E. Chicago Avenue, Chicago, IL, 60611, USA
✉ Nicholas P. Cianciotto, Northwestern University, Department of Microbiology and Immunology, 303 E. Chicago Avenue, Chicago, IL, 60611, USA (*phone* +01-312-503-1034;
fax +01-312-503-1339; *e-mail* n-cianciotto@nwu.edu)

1994), ColE1 derivatives (Pope et al, 1994) and R100 (IncFII group plasmid) derivatives (Abu Kwaik et al, 1997). Electroporation has also been used to introduce plasmids into *L. micdadei*. (O'Connell et al, 1995). The procedure presented here is derived from Cianciotto and Fields, 1992.

Outline

The technique essentially involves preparing the strains for electroporation by extensive washes in 10% glycerol. During these washes, the cells are concentrated to about 10^{11}cells/ml. These cells are frozen at -70°C, and thawed just prior to electroporation. Transformation is achieved by adding the cells to the DNA and subsequently applying a high-voltage exponential decay pulse in an electroporation chamber. Following outgrowth in Buffered Yeast Extract (BYE) broth, the transformants are selected by plating on Buffered Charcoal Yeast Extract (BCYE) agar plates containing the appropriate antibiotic.

Materials

Solutions and media (preparation and storage)

– Washing and freezing medium:
   – 10% Glycerol in sterile deionized water, autoclaved.

Note: Laboratory grade of glycerol should be sufficent.

   – Store at 4°C.
– BYE medium for outgrowth:
   – Dissolve 10g ACES, 2.8 g KOH, 10g Yeast extract, 1g α-ketoglutarate mono- potassium salt, 0.4g L-cysteine and 0.25g Ferric pyrophosphate in 1 liter of deionized water. Filter sterilize
   – Store at 4°C in a dark bottle.

Note: Alternatively, the sodium salt of α-ketoglutarate may be used.

– BCYE medium for plating the cells:
   – Dissolve 10g ACES, 2.8 g KOH, 10g Yeast extract, and 1g α-ketoglutarate monopotassium salt in 900 ml of deionized

water. Adjust the pH to 6.85-6.95 using KOH. Bring the volume up to 1L with deionized water. Add 1.5g activated charcoal and 15g agar. Autoclave for 20 minutes. Cool to 50°C.

- Dissolve 0.4g of L-cysteine in about 2 ml deionized water. Dissolve 0.25g of Ferric pyrophosphate in about 2ml deionized water. Shake the tube at 37°C for 10 minutes to dissolve the ferric pyrophosphate. Filter sterilize these two solutions and add them independently to the cooled medium. Pour into petri dishes.
- The plates can be stored in plastic sleeves at 4°C for up to six months.
- Antibiotic stock solutions (1000x): Kanamycin: 25mg/ml (filter sterilize), Chloramphenicol: 3mg/ml (in ethanol); add 1 ml of the appropriate antibiotic to BCYE media (1L) that has been cooled to 50°C. Mix well before pouring into petri plates. Store at -20°C.

Bacterial strains

L. pneumophila strains that have been used in the past include Philadelphia 1 (Marra and Shuman, 1992), 130b (Cianciotto and Fields, 1992) and RI-243 (Pruckler et al., 1995).

Equipment

Electroporation apparatus and Micro-electroporation chambers (cuvettes):
The technique described here uses the "Cell-Porator" from BRL life technologies. Other electroporation apparatus (e.g. GenePulser from Biorad) and cuvettes can also be used (Marra et al, 1992, Pruckler et al, 1995).

Note: The micro-electroporation chambers from Life technologies can be cleaned with ethanol, dried, autoclaved and re-used. We have not, however, determined if there is a loss in transformation efficiency with re-used chambers.

Time considerations

L. pneumophila takes 2-3 days to grow at 37°C. Preparation of cells for electroporation takes approximately 2 hours. Cells are frozen for later use at this stage. Electroporation (including incubation of cells with DNA) takes approximately 40 minutes. Outgrowth is for 4-6 hrs. Cells are plated subsequently. Typically, transformants appear on the plates in 4-5 days.

Safety
regulations Appropriate precautions must be ensured while using the electroporation apparatus. The high voltage, high current discharges generated during electroporation are potentially lethal. Please refer to manufacturer's recommendations for safe operation of the instrument. Explosive arcing may also occur, especially with samples of high conductivity. This may result in aerosolization of all or part of the samples. *L. pneumophila* and *L. micdadei* should be handled in BL2 conditions. Care must be taken to prevent exposure of personnel to aerosolized bacteria.

Note: Presence of salt in the DNA sample can result in arcing. In a cell-porator chamber, this often results in the expulsion/aerosolization of the liquid between the electrodes. If this happens, deionize the DNA solution by passing through a Sepharose CL-6B spin column, or by extensive ethanol washes. If the DNA sample is concentrated, dilution with deionized water may be an option.

Procedure

Preparation of *L. pneumophila* cells for electroporation

The procedure described here will yield 0.5 ml of competent cells, sufficient for about 25 electro-transformations.

1. Streak out cultures of *L. pneumophila* on BCYE plates. After 2-3 days of growth at 37°C, add about 5 ml of BYE to the plates and scrape the cells into a sterile oak ridge tube. Adjust the volume with BYE such that the $OD_{660} = 1.8$ in a final volume of about 20 ml.

Note: The cells can also be grown in stationary or shaking broth cultures. Always use freshly grown/streaked cultures.

2. Centrifuge the cell suspension at 4300g (e.g. 6000 RPM in a Beckman JA20 rotor) for 10 minutes at 4°C in a pre-chilled rotor.

3. Resuspend the cell pellet in 10 ml of cold 10% sterile glycerol, and divide into two sterile 250 ml centrifuge bottles.

4. Add 250 ml of cold 10% glycerol to each bottle.

5. Centrifuge at 4300g (e.g. 5500 RPM in a Beckman JA14 rotor) for 10 minutes at 4°C. Discard the supernatant.

6. Repeat steps 4 and 5. Remove as much of the supernatant as possible. The pellet tends to be slightly diffuse after these washes.

7. Resuspend the cells in 500 µl cold 10% glycerol. This results in a suspension of ~10^{11} colony forming units(cfu)/ml.

8. Dispense in convenient aliquots in 1.5ml sterile microfuge tubes (assume 20 µl per transformation). Cells may be used immediately or flash frozen in an ethanol/dry ice bath and stored at -70°C.

Note: Once thawed, it is not advisable to re-freeze the competent cells since this can result in a drastic reduction in viability and transformation efficiency.

Electroporation

The technique for electroporation is identical for introducing plasmids for complementation or mutagenesis, as also for counter-selection.

1. Place ice in the electroporation chamber and cool the electroporation cuvettes. Thaw the cells to be electroporated on ice. Precool the DNA samples.

2. Add 1-2 µl of DNA (0.1 µg/µl) to 20 µl of competent cells in a pre-chilled 1.5 ml microcentrifuge tube. Incubate on ice for 20 minutes. For certain purposes (e.g. allelic exchange), a higher concentration of DNA (1µg/µl) is recommended.

Note: It is imperative that the DNA used is pure, and does not contain excessive salt. The DNA is preferably resuspended in deionized water. We generally isolate the plasmid DNA from one of several *E. coli* strains (HB101, DH5αF', XL1Blue etc.). We routinely use Wizard Miniprep columns from Promega Corporation to prepare DNA samples for electroporation. The use of Qiagen plasmid kit has also been reported (Abu Kwaik et al., 1997). DNA prepared by alkaline lysis (Sambrook et al., 1989) may also be used. Since the sample is retained by capillary action between the two electrodes of the cuvette, it is not possible to increase the volume of cells + DNA. Addition of glycerol to the DNA sample allows the retention of a larger volume (25μl) between the electrodes.

3. Place 2 ml of BYE medium in sterile tubes and warm to 37°C.

4. Place the cells and DNA mixture between the electrodes of the pre-cooled electroporation cuvette. Position the cuvette in the ice-filled electroporation chamber. Deliver an exponential decay pulse of 4 KV, 330 μF capacitance and high Ω.

Note: Avoid trapping air bubbles when transferring the DNA-cell mixture to the electroporation cuvette. The presence of air bubbles can result in a lower time constant, which in turn may result in a lower transformation efficiency.

Note: Recommended settings for the Biorad-Gene Pulser apparatus: 0.2 cm cuvette gap, 2.3 kV, 25μF, 100Ω, with a time constant of 2.4 msec. Other settings have also been used. (See Survey Nos. 50, 51 and 203 in Methods in Electroporation, Gene-Pulser/ *E. coli* Pulser systems from Biorad laboratories, 1993).

5. Outgrowth: Immediately after the pulse, transfer the cells to tubes containing 2 ml of pre-warmed BYE medium. Incubate with shaking at 37°C for 1-5 hours.

Note: Phenotypic expression beyond 6 hours is not advisable since cell division may occur. This will result in colonies that arise due to cell division, rather than initial transformation.

6. Plate 100 μl of the cell suspension on selective BCYE plates. If a lower efficiency of transformation is expected, pellet the cells first. Resuspend the pellet in 100 μl of BYE medium and plate on selective BCYE plates.

Note: Depending on the amount of DNA used, it may be necessary to plate dilutions of the transformed cells.

Controls   The following controls will facilitate troubleshooting problems during transformation. They are designed to determine the number of cells initially used, the number of cells surviving the electric pulse and the number of surviving cells transformed.

An additional DNA-less control is used to verify the integrity of the antibiotic plates.

1. Plate dilutions of the electroporation-competent cells on plain BCYE plates to determine the titer of cells in the sample prior to electroporation.

2. Plate dilutions of the electroporated sample on plain BCYE plates. This will give an estimate of the percent of cells that survive electroporation. We have observed 20-50% survival for the process.

3. Determine the number of transformants obtained per µg of DNA. Using the values obtained above, the number of transformants/recipient/µg of DNA can be determined. This can then be used to compare various batches of electrocompetent cells.

4. A negative control could also be included to verify the integrity of the selective plates. For this, plate 20 µl of competent cells directly on the selective plates.

Results

The 15-kb plasmid pNC31.3 plasmid was transferred into the strain NU201 (a spontaneous streptomycin-resistant derivative of the virulent clinical isolate, 130b) at a frequency of about 10^{-3} Km^r cfu/recipient or 4 x 10^5 Km^r cfu/µg of DNA using the above protocol. In contrast, electroporation of strain 130b with the 8.6 kb Km^r plasmid pEYDG1 yielded 10^{-2} Km^r cfu/recipient and resulted in ~1 x 10^6 Km^r cfu/µg of DNA, suggesting that electroporation of smaller plasmids is more efficient (Cianciotto and Fields, 1992).

The efficiency of electroporation into *L. micdadei* was significantly lower than it was for *L. pneumophila* i.e., ca 10^2 versus 10^5 cfu/µg of DNA. The procedure was used to introduce a ColE1 derivative containing a counter-selectable marker into *L. micdadei* (O'Connell et al, 1995).

In general, differences between strains and species can result in large fluctuations in transformation efficiency. To increase transformation efficiency, it may be necessary to adapt the method to the particular strain in use. Care must be taken, however, to avoid excessive cell killing (Miller F, 1994).

References

Abu Kwaik Y, Pederson LL (1996) The use of differential display-PCR to isolate and characterize a *Legionella pneumophila* locus induced during the intracellular infection of macrophages. Mol Microbiol 21(3):543-56.

Abu Kwaik Y, Gao LY, Harb OS, Stone BJ (1997) Transcriptional regulation of the macrophage-induced gene (*gspA*) of *Legionella pneumophila* and phenotypic characterization of a null mutant. Mol Microbiol 24(3):629-42.

Berger KH, Isberg RR (1993) Two distinct defects in intracellular growth complemented by a single genetic locus in *Legionella pneumophila*. Mol Microbiol 7(1):7-19.

Berger KH, Merriam JJ, Isberg RR (1994) Altered intracellular targeting properties associated with mutations in the *Legionella pneumophila dotA* gene. Mol Microbiol 14(4):809-22.

Brand BC, Sadosky AB, Shuman HA (1994) The *Legionella pneumophila icm* locus: a set of genes required for intracellular multiplication in human macrophages. Mol Microbiol 14(4):797-808.

Cianciotto NP, Eisenstein BI, Engleberg NC, Shuman H (1989) Genetics and Molecular Pathogenesis of *Legionella pneumophila* an intracellular parasite of macrophages Mol Biol Med 6:409-424.

Cianciotto NP, Fields BS (1992) *Legionella pneumophila mip* gene potentiates intracellular infection of protozoa and human macrophages, Proc Natl Acad Sci USA, 89:5188-5191.

Hickey EK, Cianciotto NP (1997) An iron- and *fur*-repressed *Legionella pneumophila* gene that promotes intracellular infection and encodes a protein with similarity to the *Escherichia coli* aerobactin synthetases. Infect Immun 65(1):133-43.

Marra A, Blander SJ, Horwitz MA, Shuman HA (1992) Identification of a *Legionella pneumophila* locus required for intracellular multiplication in human macrophages Proc Natl Acad Sci USA. 89:9607-9611.

Marra A, Shuman HA (1992) Genetics of *Legionella pneumophila* virulence. Annu Rev Genet 26:51-69.

McClain MS, Engleberg NC (1996) Construction of an alkaline phosphatase fusion-generating transposon, mTn*10phoA*. Gene 170:147-148.

Miller F (1994) Bacterial transformation by electroporation. Methods in Enzymology, vol 235. Academic Press, London, pp 375-385.

O'Connell WA, Bangsborg JM, Cianciotto NP (1995) Characterization of a *Legionella micdadei mip* mutant. Infect and Immun 63(8):2840-2845.

Pope CD, Dhand L, Cianciotto NP (1994) Random mutagenesis of *Legionella pneumophila* with mini-Tn*10*. FEMS Microbiol Lett 124(1):107-11.

Pruckler JM, Benson RF, Moyenuddin M, Martin WT, Fields BS (1995) Association of flagellum expression and intracellular growth of *Legionella pneumophila*. Infect Immun 63(12) 4928-4932.

Sadosky BA, Wiater LA, Shuman, HA (1993) Identification of *Legionella pneumophila* genes required for growth within and killing of human macrophages. Infect Immun 61(12):5361-5373.

Sambrook J., Fritsch EF, Maniatis T. (1989) Molecular Cloning: A laboratory manual. Cold Spring Harbor Laboratory, New York, Cold Spring Harbor Laboratory Press.

Wiater LA, Sadosky BA, Shuman (1994) Mutagenesis of *Legionella pneumophila* using Tn*903*dII*lacZ*: identification of a growth-phase-regulated pigmentation gene. Mol Microbiol 11(4):641-653.

Electrotransformation
of *Streptococcus pneumoniae*

JACQUES LEFRANCOIS and ARMAND MICHEL SICARD

Introduction

Biotechnology has undergone rapid development as a result of the discovery of transformation methods. The first bacteria naturally able to be transformed by DNA, was *Streptococcus pneumoniae*. However for decades, it was difficult to obtain competence. The complex culture media were not reliable and failure to obtain transformation for unknown reasons could last for weeks or months. More reproducible procedures progressively devised by a few laboratories improved the efficiency of natural transformation. Only very recently were two 17-aminoacids oligopeptides that induce competence in pneumococcus discovered and isolated (Havarstein et al., 1995; Pozzi et al., 1996). These synthetic pheromones greatly facilitate transformation experiments and have been tested on several capsular types. They induced competence at a variable level according to the strains tested. However 50% of these strains could not be transformed using these pheromones. As it is necessary to use a universal procedure to obtain DNA induced transformation able to be applied to many strains, we have decided to use electroporation in *Streptococcus pneumoniae*. Several years ago, we obtained transformants induced by an electric field in a laboratory strain, R36A, a rough derivative of a type II pneumococcus (Bonnassie et al., 1989). Electrotransformation was successful for capsulated strains and for strains deficient in their natural competence.

✉ Jacques Lefrancois, LMGM du CNRS, 118 route de Narbonne, Toulouse Cedex, 31062, France (*phone* 33-5-61 33 58 00; *fax* 33-5-61 33 58 86; *e-mail* lfrancois@ipbs.fr)
Armand Michel Sicard, LMGM du CNRS, 118 route de Narbonne, Toulouse Cedex, 31062, France

Moreover we investigated the process of DNA uptake during electrotransformation as compared with natural competence. It has been found that in contrast to natural transformation double-stranded DNA penetrates in *Streptococcus pneumoniae* without a single-stranded DNA step (Lefrançois and Sicard, 1997). Single-stranded by itself is very weakly transformant and linearized DNA plasmids yield barely detectable levels of transformants. Chromosomal markers cannot be transformed by electric treatment (Lefrançois et al., 1998). Moreover DNA is strongly restricted by the DpnI or DpnII pneumococcal restriction enzymes when introduced by electrotransformation. During these studies the electrotransformation protocole has been improved and is presented in the following sections. It was found that several factors affect the frequency of transformation : only early logarithmic phase cultures should be used and addition of serum albumin during the culture increases transformation frequency tenfold. Routinely the frequency of transformants is 10^{-3} per µg of plasmid DNA ml^{-1}.

Materials

Electrotransformation was performed using the Bio-Rad Gene Pulser. When needed, a Pulse Controller apparatus was also used. Cells were washed and concentrated using a Sorvall RC 5B centrifuge. Plasmids were purified by CsCl ethidium bromide density-gradient in a Beckman Model L 8-70 ultracentrifuge.

Equipment

Antibiotics included Erythromycin (Sigma), Streptomycin (Sigma), Tetracyclin (Sigma). Fraction V bovalbumin was obtained from Sigma or Euromedex. Platings were made in plastic petri dishes (90mm, Phoenix Biomedical) containing agar-growth medium and defibrinated blood (bioMérieux).

Antibiotics

Cultures were grown at 37°C without aeration from an 1/100 inoculum. Cells were routinely harvested in early - logarithmic phase (OD$_{550}$ 0,15 - 0,20). Cells were grown in complete CAT broth (Modified from Morrison et al., 1983). This medium contains per liter : 5 g of Tryptone (Difco), 10 g of Casitone (Difco), 1 g yeast extract (Difco) and 5 g of NaCl (Merck). It was sterilized for 20 min. at 120°C and then brought to 0.2 % glucose and 1/60

Culture media

(0,0167 M) dipotassium phosphate before use. Albumin (fraction V) solutions were prepared at a concentration of 8 percent and ajusted to pH 2,5 with HCl 35% rectapur Prolabo. They were sterilized under shaking at 100°C for 30 min., and stored acid, in the cold. Prior to use, they were slowly neutralized (pH 7,5) with NaOH 3N allowing constant shaking. CAT medium was supplemented with 0,2 % neutralized albumin before being used for growth.

Electroporation buffer: Cells were electroporated in a phosphate, sucrose, magnesium buffer (PSM): 7 mM-potassium phosphate pH 7.5, 0.5 M sucrose, 1 mM $MgCl_2$.

Viable cells and selection of transformants

Cells   Cultures were diluted in CAT medium and suitable aliquots were plated in melted complete agar medium (gelose D) kept at 55°C (0.1% Difco dextrose, 0.5% Difco neopeptone, 0.125% Difco yeast extract, 0.125% trizma base from Sigma 0.5% NaCl from Merck,1% Bio-Trypcase from bioMérieux 1% Oxod Bacto, pH 7.2) into which is incorporated at the time of plating 3.3% of defibrinated horse blood and the dilution of bacteria. The mixture is homogenized by a gentle shaking of each plate and kept on the bench. When solidified they are transferred in an incubator at 36-37°C for 24 hours. Transformants were scored on the same medium. However to allow phenotypic expression of antibiotic resistance, the plates containing 10 ml of blood, complete agar and aliquots of culture were incubated 2,5 hours at 37°C and received 10 ml of an overlayer of the same complete agar medium without blood and containing the suitable antibiotic at a concentration twice the resistance level conferred by each marker. Plasmid installation was routinely selected at a final concentration of 3 µg/ml erythromycin or 1,5 µg tetracycline.

Strains and plasmids   *Streptococcus pneumoniae* strain R801 or R800 used in this study were derived from Avery's strain R36 A. The plasmids were either pLS1 that replicates in *Escherichia coli* and *S. pneumoniae* (Stassi et al., 1981) which carries a resistance marker for 1,5 µg ml^{-1} tetracycline or pSP2 which carries the same resistance marker and a marker conferring resistance to 4 µg / ml^{-1} erythromycin (Prats et al., 1985).

Procedure

Culturing *Streptococcus pneumoniae*

Cultures were grown at 37°C without aeration in the complete CAT-broth containing 0.2% neutralized albumin. They were inoculated by a dilution 1/50 or 1/100 of a preculture kept frozen at -70°C on 15% glycerol in the same medium. Cells were harvested in early logarithmic phase (OD_{550} 0,15 - 0,20). A 800 ml culture was centrifuged for 15 min. at 4°C and 5000 rpm in the Sorvall RC 5B centrifuge. The pellets were resuspended in 80 ml of CAT - albumin - medium. They can be kept, at -70°C for several months when glycerol has been added (15%).

Prior to electrotransformation, frozen cultures were melted at room temperature and kept in ice. Cells were washed twice in the electroporation medium and concentrated five fold in the same buffer. A volume of 0,8 ml of this cell suspension was poured in the 0,4 cm Bio-Rad cuvette. The plasmid DNA was added at a concentration of 0.3 µg / ml or more. The mixture was mixed and kept in ice for one minute. The setting of the apparatus was at its maximum (2.5 KV, 25µF, no pulse controller). One single impulsion was given. Under such conditions time-constant was 4,0 to 4,6 ms. An aliquot of 0,1 ml or serial dilutions were plated on blood agar medium (gelose D). Electrotransformants were scored directly from 0.1 ml aliquots or from 10^{-1} 10^{-2} dilutions. Total cell count was obtained from 10^{-5} and 10^{-6} dilutions. Viable cells were scored before and after the electric pulse. Best electrotransformation results are obtained for 70% - 90% cell survival. **Electrotransformation**

An alternative protocol has been used for refractory strains especially when their cell - wall was modified (Séroude et al., 1993). The setting of the apparatus is at its maximum and the pulse controller between 200 and 600 ohms, resulting in an electric field of 12.5KV.cm^{-1} in a 2 mm wide electroporation cuvette. To avoid electric arcs, cells were washed and resuspended in a non-ionic 0.5 M sucrose buffer ; 0,3 ml of this suspension was poured into the cuvette. Under these conditions survival might be reduced to 10% with a 200 ohm setting required to obtain transformants for some strains.

Transformation efficiency

The transformation frequency is the number of transformants divided by the number of total viable cells. This frequency is proportional to the concentration of plasmid DNA (Bonnassie et al. 1989). Therefore it is possible to express the efficiency of transformation referring to the amount of added DNA i.e. transformation efficiency per microgram of DNA / ml.

Requirements for electrotransformation

Early phase of growth Cells were collected at different growth phases and tested for electrotransformation. Efficiency of transformation was 2.10^{-3} when cells were harvested at OD_{500} 0,20. Electrotransformants could not be obtained at late growth phase or early stationary phase at OD_{500} 0,80.

Effect of albumin in the culture medium When albumin was omitted from the culture medium the efficiency of electrotransformation was reduced ten-fold. Addition of albumin in the electrotransformation buffer did not improve the efficiency. Although the role of albumin in the growth medium is obscure, it might protect the cells from some toxic agents contained in complete media as it has been hypothesized to explain that albumin is also required to obtain natural competence in *Streptococcus pneumoniae*.

Sensitivity to restriction Occurrence of a restriction barrier in electroporation has been reported for some bacteria whereas in other bacteria the electroporated plasmids can escape the restriction systems present in the recipient cells. In *Streptococcus pneumoniae* two restriction systems have been well characterized. All strains tested except a null mutant belong to DpnI or DpnII restriction system : DpnI strains do not methylate adenine in GATC sequences and restrict DNA when these sequences are methylated, whereas DpnII strains methylate adenine in this sequence and do not restrict non methylated DNA (Vovis et Lacks, 1977). Using these panels of strains we have found that electrotransformation efficiency was 2.10^{-3} per μg ml^{-1} for non restricted DNA and undetectable ($< 10^{-8}$) for restricted DNA (Lefrançois and Sicard, 1997). Therefore plasmids should be prepared in the strain to be electrotransformed that shares the same restriction condition or tested in a Dpn null strain (CP1000).

Electrotransformants can only be obtained by plasmid DNA able to replicate in *Streptococcus pneumoniae*. Chromosomal markers cannot be transformed by the electric pulse (Bonnassie, Intercept Ltd.) in contrast with natural transformation (Lefrançois et al., 1998). This failure to electrotransform chromosomal markers has not been elucidated yet. Linearized plasmid DNA yields barely detectable electrotransformant. Denaturation reduced ten fold the efficiency of transformation.

Molecular state of DNA

Comments

Electrotransformation is an efficient method to introduce replicative plasmids in *Streptococcus pneumoniae*. The protocol is simple, the media can be easily prepared and reproducibility is high if the described care has been taken, especially a rapid growth and electrotransformation on early log-phase cultures. For refractory strains, higher electric field might be useful. Sensitization of these cells by a mild penicillin treatment, can also be tried, as reported for other bacteria (Bonnassie et al., 1990).

References

Bonnassie S, Gasc AM, Sicard AM (1989) Transformation by electroporation of two Gram-positive Bacteria *Streptococcus pneumoniae* and *Brevibacterium lactofermentum*. Genetic Transformation and Expression. Intercept Limited.

Bonnassie S, Burini JF, Oreglia J, Trautwetter A, Patte JC, Sicard AM (1990) Transfer of plasmid DNA to *Brevibacterium lactofermentum* by electrotransformation. Journal of General Microbiology 136 : 2107-2112.

Havarstein LS, Coomaraswamy G, Morrison DA (1995) An unmodified heptapeptide pheronome induced competence for genetic transformation in *Streptococcus pneumoniae*. Proc Natl Acad Sci USA 92:11140-11144

Lefrançois J, Sicard AM (1997) Electrotransformation of *Streptococcus pneumoniae* : evidence for restriction of DNA on entry. Microbiology 143 : 523-526

Lefrançois J, Samrakandi M, Sicard AM (1998) electrotransformation and natural transformation of *Streptococcus pneumoniae* : requirement of DNA processing for recombination. Microbiology 144: 3061-3068

Pozzi G, Masala L, Iannelli F, Manganelli R, Havarstein LS, Piccoli L, Simon D; Morrison DA (1996)Competence for genetic transformation in encapsulated strains of Streptococcus pneumoniae: two allelic variants of the peptide pheromone. J Bacteriol 178: 6087-90

Prats H, Martin B, Pognonec P, Burger AC, Claverys JP (1985) A plasmid vector allowing positive selection of recombinant plasmids in *Streptococcus pneumoniae*. Gene 39:41-48

Seroude L, Hespert S, Selakovitch-Chenu L, Gasc AM, Lefrançois J, Sicard M (1993) Genetic studies of cefotaxime resistance in *Streptococcus pneumoniae* : relationship to transformation deficiency. Res Microbiol 144 : 389-394

Stassi D, Lopez P, Espinoza M, Lacks SA (1981) Cloning of chromosomal genes in *Streptococcus pneumoniae*. Proc Natl Acad Sci 78: 7028-7032

Morrison, D.A., Lacks, S.A., Guild, W.R. & Hageman, J.M. 1983. Isolation and characterization of three new classes of transformation deficient mutants of *Streptococcus pneumoniae* that are defective in DNA transport and genetic recombination. J. Bacteriol. 156 : 281-290.

Vovis, G.F. & Lacks, S.A. 1977. Complementary action of restriction enzymes endo R. DpnI and endo R. DpnII on bacteriophage f1. J. Mol. Biol. 115 : 525-538.

Part IV

Plants

Clavibacter michiganensis - Transformation of a Phytopathogenic Gram-Positive Bacterium

DIETMAR MELETZUS, HOLGER JAHR and RUDOLF EICHENLAUB

Introduction

The Gram-positive bacterium *Clavibacter michiganensis* subsp. *michiganensis* (Smith) causes bacterial wilt and canker of tomato (*Lycopersicon esculentum* Mill.). The bacteria infect the host plant via wounds, invade the xylem vessels followed by a systemic infection of the host. This pathogen is the target of international quarantine regulations. Bacterial canker of tomato has caused major economic losses of up to 60 % in commercial tomato production world-wide, since neither resistant tomato cultivars nor effective chemical controls of this pathogen are available. *C. m. michiganensis* generally is transmitted by contaminated seeds or transplants, and therefore the spreading of this serious pathogen can only be controlled by detection and elimination of infected plants.

For a genetic approach to the mechanisms of pathogenicity a vector and transformation system was developed for *C. m. michiganensis* (Meletzus and Eichenlaub 1991, Laine et al. 1996). The involvement of plasmids pCM1 and pCM2 in pathogenicity along with the localization of plasmid encoded pathogenic determinants and an endocellulase was reported recently (Meletzus et al. 1993, Dreier et al. 1997).

Dietmar Meletzus, Universtät Bielefeld, Fakultät für Biologie, Lehrstuhl Mikrobiologie / Gentechnologie, Bielefeld, 33501, Germany
Holger Jahr, Universtät Bielefeld, Fakultät für Biologie, Lehrstuhl Mikrobiologie / Gentechnologie, Bielefeld, 33501, Germany
✉ Rudolf Eichenlaub, Universtät Bielefeld, Fakultät für Biologie, Lehrstuhl Mikrobiologie / Gentechnologie, Bielefeld, 33501, Germany
(*phone* +49-521-106-5558; *fax* +049-521-106-6015;
e-mail Eichenlaub@biologie.uni-bielefeld.de)

Although DNA uptake was demonstrated by PEG-mediated spheroplast transformation, satisfying transformation results were only obtained by electroporation of intact or lyzozyme-treated *Clavibacter* cells.

Outline

Late logarithmic cells are washed to remove the growth medium, cells are converted to spheroplasts by a mild lysozyme treatment, concentrated by centrifugation and then used for electrotransformation.

Materials

Equipment
- Gene Pulser Apparatus (Bio-Rad), Pulse Controller (Bio-Rad)
- Electroporation cuvettes (0.2 cm) (Eurogentec)

Buffers
- Washing buffer: 10 % (v/v) Glycerol in deionized H_2O
- Protoplasting buffer: 40 µg Lysozyme (Sigma, L-6876) per ml in washing buffer

Media
- Growth medium (C-medium)
  - 10 g peptone
  - 5 g yeast extract
  - 5 g glucose
  - 5 g NaCl
  - add. 1000 ml H_2O, pH 7.2
- Regeneration medium A
  - 10 g tryptone
  - 5 g yeast extract
  - 4 g NaCl, pH 7.2
  - add 500 ml deionized H_2O, autoclave.
- Regeneration medium B: 460 ml 1 M sorbitol, pH 7.2
- Regeneration medium
  - 500 ml medium A
  - 460 ml medium B
  - 20 ml 1 M $MgCl_2$
  - 20 ml 1 M $CaCl_2$

– For a solid regeneration medium add agar to a final concentration of 1.5 % (w/v)

Procedure

1. Grow 100 ml *Clavibacter* cells to late logarithmic growth-phase (approx. 1 x 10^9 cells per ml).

   Preparation of cells

2. Collect cells by centrifugation at 3000 x g for 10 min at 4°C.

Note: All of the following steps are carried out on ice or at 4°C as indicated.

3. Wash cells by resuspension of the bacterial pellet in 10 ml of H_2O (0°C), followed by centrifugation at 3000 x g for 10 min at 4°C.

4. Resuspend the bacterial pellet in 10 ml of Protoplasting Buffer.

5. Inoculate for 20 min at 37°C.

6. Collect cells by centrifugation at 3000 x g for 15 min at 4°C.

7. Wash cells once in 10 ml of washing buffer (0°C).

8. Collect cells by centrifugation at 3000 x g for 15 min at 4°C.

9. Resuspend the bacterial pellet in 2.5 ml of washing buffer (0°C).

1. Add up to 4 μg of supercoiled plasmid DNA to 200 μl spheroplasted cells.

   Electroporation

2. Transfer DNA-cell mixture to a precooled electroporation cuvette.

3. Apply appropriate pulse.

4. Add 200 μl of regeneration medium and mix carefully.

5. Incubate cells at 25°C for 3 hrs.

6. Plate on solid regeneration medium containing the appropriate antibiotics.

Electroporation parameters

- Adjust Gene Pulser apparatus to a field strength of 12.5 kV per cm.
- Adjust Pulse Controller to a resistance of 600 Ω.
- Pulse length should be in the range of 11.5 to 13.5 ms.

Results

When host specific modified DNA was used, transformation rates were in the range of 2×10^3 transformants per μg of plasmid DNA for *Clavibacter michiganensis* subsp. michiganensis (Meletzus and Eichenlaub 1991). Using plasmid DNA isolated from *E. coli*, the transformation efficiency dropped by a factor of 100, possibly due to an active restriction-modification system. To collect some information on the general usefulness of this method as a transformation procedure for related *Clavibacter* strains, we were able to demonstrate successful electroporation of *Clavibacter michiganensis* subsp. *sepedonicus, Clavibacter michiganensis* subsp. *nebraskensis, Clavibacter michiganensis* subsp. *insidiosium,* and *Clavibacter iranicus.*

A slight modification of the method described here, using 0.1 cm cuvettes, a field strength of 18 kV/cm, 600 Ω, and a pulse duration from 10.7 to 12.6 ms resulted in a maximum yield of 4.6×10^4 transformed cells per μg of plasmid DNA for *Clavibacter michiganensis* subsp. *sepedonicus* (Laine et al. 1996).

Troubleshooting

- If the pulse length observed is shorter than 11,5 to 13.5 ms, dilute the bacterial cell suspension by adding an appropriate volume of washing buffer.
  Alternatively use the pulse controller unit to adjust the appropriate pulse length.

- Highly purified DNA-preparations are crucial to avoid high conductivity.
  Preferably the isolations should be carried out by the affinity column technique (Qiagen).

- DNA volumes added should not exceed 5 μl.
  For resuspension of DNA after precipitation ultrapure H_2O is
  recommended.

Comments

The method described above has been used successfully for the
cloning of virulence genes of both, *Clavibacter michiganensis*
subsp. *michiganensis* (Dreier et al. 1997) and *Clavibacter michiganensis* subsp. *sepedonicus* (Laine et al. 1996), respectively.
Transformation of these bacteria is most effective using supercoiled plasmid DNA. The use of linear DNA in *Clavibacter michiganensis* subsp. *michiganensis* electroporation experiments
has not been successful so far.

Acknowledgements. This work was supported by the Deutsche Forschungsge-
meinschaft.

References

Dreier,J., Meletzus,D., Eichenlaub,R. (1997) Characterization of the Plasmid
    encoded virulence region *pat*-1 of phytopathogenic *Clavibacter michi-
    ganensis* subsp. *michiganensis*. MPMI 10(2):195-206
Meletzus D., Eichenlaub R. (1991) Transformation of the phytopathogenic
    bacterium *Clavibacter michiganense* subsp. *michiganense* by electro-
    poration and development of a cloning vector. J. Bacteriol.173:184-190
Meletzus D., Bermpohl A., Dreier J., Eichenlaub R. (1993) Evidence for plas-
    mid encoded virulence factors in the phytopathogenic bacterium *Clavi-
    bacter michiganense* subsp. *michiganense* NCPPB382 J. Bacteriol.
    175:2131-2136
Laine MJ, Nakhei H, Dreier J, Lehtilä K, Meletzus D, Eichenlaub R, Metzler
    MC (1996) Stable transformation of the gram-positive phytopathogenic
    bacterium *Clavibacter michiganensis* subsp. *sepedonicus* with several
    cloning vectors Appl. Env. Microbiol. 62: 1500-1506.
Yoshihama M, Higashiro K, Rao EA, Akedo M, Shanabruch WG, Folletie
    MT, Walker GC, Sinskey AJ (1985) Cloning vector sytem for *Corynebac-
    terium glutamicum*, J. Bacteriol. 162: 591-597.

Suppliers

SIGMA: Grünwalder Weg 30, D-82041 Deisenhofen

BIO-RAD: Heidemannstraße 164, D-80939 München

EUROGENTEC: Parc scientifique du Sart Tilman, 4102-Belgium

QIAGEN GmbH: Max-Volmer-Str. 4, D-40724 Hilden

Electrotransformation of *Agrobacterium tumefaciens* and *A. rhizogenes*

DIETHARD MATTANOVICH and FLORIAN RÜKER

Introduction

Agrobacterium tumefaciens is routinely used to transfer DNA into plants (e.g. reviewed by Hooykaas 1989). Naturally, this process is brought about by a class of plasmids called Ti (tumour inducing). Similarly, *A. rhizogenes* strains contain an Ri (root inducing) plasmid. Parts of these functions, deprived of the tumour inducing genes (which code for plant hormone synthesising enzymes), are used for plant transformation. Classically, mating procedures with helper strains have been utilized to transfer plasmids carrying the recombinant plant genes into *Agrobacterium* (Ditta et al. 1980). Direct transfer of plasmid DNA has been attempted yielding low or no results (Holsters et al. 1978). In 1989 we developed an electrotransformation procedure for different strains of *A. tumefaciens* and *A. rhizogenes* (Mattanovich et al. 1989). Basically this procedure resembles most of the protocols used for gram negative bacteria (Dower et al. 1992). The major advantage of this procedure is a significant reduction of time and labour as compared to the mating procedure.

The genus *Agrobacterium* (Rhizobiaceae) consists of four species: *A. tumefaciens, A. radiobacter, A. rhizogenes* and *A. rubi* (Holt et al. 1994). Phenotypically *Agrobacterium* strains are classified in three biovars according to their nutritional properties. In our work we used two strains of *A. tumefaciens* (allo-

✉ Diethard Mattanovich, Universität für Bodenkultur, Institut für Angewandte Mikrobiologie, Muthgasse 18, Vienna, 1190, Austria (*phone* +43-1-36006-6569; *fax* +43-1-3697615; *e-mail* d.mattanovich@iam.boku.ac.at)
Florian Rüker, Universität für Bodenkultur, Institut für Angewandte Mikrobiologie, Muthgasse 18, Vienna, 1190, Austria

cated to biovar 1) and one strain of *A. rhizogenes* (biovar 2) with similar results.

Materials

Equipment   Electroporator (e.g. Bio-Rad Gene Pulser)

Although not determined experimentally, we assume that any instrument which is suitable for electrotransformation of *Escherichia coli* can be used instead.

Growth media   LB glc medium (for biovar 1 strains)

bacto-tryptone	10 g/l
yeast extract	5 g/l
NaCl	5 g/l
glucose	5 g/l

Adjust pH to 7.0 with HCl or NaOH and autoclave.

YMB medium (for biovar 2 strains)

K_2HPO_4	0.5 g/l
NaCl	0.1 g/l
Yeast extract	0.4 g/l
Mannitol	10 g/l

1. Adjust pH to 7.0 with HCl, then adjust volume to 990 ml and autoclave.

2. Dissolve 2 g $MgSO_4$ in 10 ml, autoclave separately, add after cooling.

SOC medium

Bacto-tryptone	20 g
Yeast extract	5 g
NaCl	0.5 g

1. Dissolve in 950 ml distilled water.

2. Add 10 ml of 250 mM KCl.

3. Adjust pH to 7.0 with 5 N NaOH.

4. Adjust volume to 975 ml, autoclave.

5. After cooling, add 20 ml 1 M glucose and 5 ml 2 M $MgCl_2$.

Buffers and
solutions

- HEPES buffer
  - 1 M HEPES
  - Adjust pH to 7.0, filter sterilize.
  - Dilute 1:1000 with sterile distilled water before use.
- Glycerol solution
  - 10 % (w/v) Glycerol
  - Sterilize by autoclaving.

Procedure

Preparing competent cells

1. Inoculate 300 ml LB Glc medium (or YMB resp.) in a 1 l   **Bacterial**
   baffled shake flask with 3 ml of an overnight culture.   **culture**

2. Shake at 28°C until OD_{600} reaches 0.5 (this step takes 5-8 h).

3. Centrifuge at 3000 × g for 10 min at 4°C, discard supernatant.   **Harvest**

4. All solutions and the culture must be kept cold throughout
   the following procedures.

5. Resuspend in 250 ml 1 mM HEPES buffer, centrifuge as in   **Washing**
   step 3.

6. Repeat step 5 in 200 ml 1 mM HEPES buffer.

7. Repeat step 5 in 100 ml 1 mM HEPES buffer.

8. Repeat step 5 in 30 ml 10 % glycerol.

9. Resuspend in 1 ml 10 % glycerol, aliquot at 100 μl.

10. Quick-freeze in liquid nitrogen or -70°C methanol, store at   **Storages**
    -70°C. Shelf life is at least 6 months.

Transformation

1. Thaw competent bacteria slowly on ice.   **Competent
cells**

2. Add 100 ng plasmid DNA (dissolved in water).   **Plasmid DNA**

3. Keep on ice for 5 minutes.

4. Transfer into electroporation cuvette: 100 µl for 4 mm cuvettes, 40 µl for 2 mm cuvettes.

Electrotrans-
formation

5. Standard settings on electroporator (Bio-Rad Gene Pulser): 2.5 kV; 25 µF; pulse controller at 1000 Ohm.

6. Apply 1 pulse.

7. Immediately add 1 ml SOC medium (or YMB resp.), transfer into sterile 10 ml tube.

8. Incubate with shaking at 28°C for 1-3 hours.

Selection

9. Streak out 100 µl on an LB Glc (YMB resp.) agar plate containing the appropriate selective antibiotic.

10. Incubate at 28°C for two days.

Results

Different strains of *A. tumefaciens* and *A. rhizogenes* have been successfully transformed using this procedure. The highest yields that have been achieved with *A. tumefaciens* strain LBA4404 and the binary plasmid pDG12Sa were 1.5×10^6 transformants per µg plasmid DNA. Using other strains, Mersereau et al. (1990) have achived around 10^8 transformants per µg. It can be assumed that the electrotransformation efficiency is strain dependent, as it is for *E. coli*.

As shown in Table 1, both the initial electrical field strength and the time constant of the pulse (determined by the resistance setting of the pulse controller) have a significant influence on the yield of transformation.

Table 1. Electrotransformation yields of *A. tumefaciens* strain LBA4404 (Hoekema et al. 1983) with plasmid pDGSa12 (Vilaine and Casse-Delbart 1987), comparing different electroporation cuvettes and different resistance settings on the Bio-Rad Gene Pulser

Electrode gap (mm)	Field strength (kV/cm)	Resistance (Ohm)	Yield (Colonies/µg)
4	6.25	200	3.8×10^3
4	6.25	1000	3.4×10^4
2	12.5	200	2.3×10^4
2	12.5	1000	5.5×10^5

However, it should be noted that in the experiments described here both parameters have been set to the maximum values allowed by the apparatus without exceeding a distinct maximum of transformation yield, so that an optimum yield cannot clearly be identified.

Comments

Critical points

- Agrobacteria should preferably be grown in baffled shake flasks using a rotary shaker at around 200 rpm.
- Preheating the growth media is recommended.
- Adding glucose to LB medium significantly promotes growth.
- After harvesting, the bacterial culture and all solutions must be kept cold during the preparation of competent cells.
- Prolonged incubation (3 hours rather than 1) in growth medium after electroporation has been shown to improve the yield by a factor of 2 (Lin 1994)

Cultivation of bacteria

As shown in Table 1, electroporation cuvettes with 2 mm gap give better results than home-made 4 mm cuvettes. However, the use of home-made cuvettes (photometer cuvettes with self-adhesive aluminium foil glued to the side walls) is by far the more economic solution for routine work, where high yields are not so relevant. The maximum yields of transformation would allow, on the other hand, for direct cloning in *Agrobacterium*, e.g. for the generation of gene libraries which could be subsequently transferred into plants.

Type of cuvette

Outlook

Preliminary work in our laboratory has indicated an additional optimization potential for electrotransformation with higher initial voltages. However, the configuration of the electrodes as used by us and most other groups limits this approach due to the occurrence of spark discharge through the air above the sample suspension. Spark discharge causes heat and deviates most of

Avoiding spark discharge

the voltage from the sample, so that transformation does not take place. It occurs above a critical field strength, depending on the isolating properties of the medium between the electrodes. As air is a much weaker isolator than water, it is by far the most limiting factor for the field strength, and replacing the air by a strong isolator should effectively prevent spark discharge.

Acknowledgements. The authors wish to thank Eva Obermayr for expert technical assistance.

References

Ditta G, Stanfield S, Corbin D, Helinski DR (1980) Broad host range DNA cloning system for gram-negative bacteria: construction of a gene bank of *Rhizobium meliloti*. Proc Natl Acad Sci USA 77:7347-7351

Dower WJ, Chassy BM, Trevors JT, Blaschek HP (1992) Protocols for transformation of bacteria by electroporation. In: Chang DC, Chassy BM, Saunders JA, Sowers AE (eds) Guide to electroporation and electrofusion. Academic Press, San Diego, pp 485-499

Hoekema A, Hirsch PR, Hooykaas PJJ, Schilperoort RA (1983) A binary plant vector strategy based on separation of *vir*- and T-region of *Agrobacterium tumefaciens* Ti-plasmid. Nature 303:179-180

Holsters M, de Waele D, Depicker A, Messens E, van Montagu M, Schell J (1978) Transfection and transformation of *Agrobacterium tumefaciens*. Mol Gen Genet 163:181-187

Holt JG, Krieg NR Sneath PHA, Staley JT, Williams ST (eds) (1994) Bergey's manual of determinative bacteriology, 9th ed, Williams & Wilkins, Baltimore

Hooykaas PJ (1989) Transformation of plant cells via *Agrobacterium*. Plant Mol Biol 13:327-336

Lin JJ (1994) Optimization of the transformation efficiency of *Agrobacterium tumefaciens* cells using electroporation. Plant Sci 101:11-15

Mattanovich D, Rüker F, da Cmara Machado A, Laimer M, Regner F, Steinkellner H, Himmler G, Katinger H (1989) Efficient transformation of *Agrobacterium* spp. by electroporation. Nucleic Acids Res 17:6747

Mersereau M, Pazour GJ, Das A (1990) Efficient transformation of *Agrobacterium tumefaciens* by electroporation. Gene 90:149-151

Vilaine F, Casse-Delbart F (1987) A new vector derived from *Agrobacterium rhizogenes* plasmids: a micro-Ri plasmid and its use to construct a mini-Ri plasmid. Gene 55:105-114

Part V

Environmental Bacteria

Transformation of the Filamentous Cyanobacterium *Fremyella diplosiphon*

ARTHUR R. GROSSMAN and DAVID M. KEHOE

Introduction

One of the most thoroughly studied signal transduction pathways in cyanobacteria controls a process known as complementary chromatic adaptation (CCA). During CCA, cells sense the spectral distribution of ambient light and adjust the pigment-protein composition of their photosynthetic light harvesting antennae structure, which is termed a phycobilisome, to maximize light absorbance. Thus, the pigmentation of the cells is linked to the spectral quality of the light in which the cells are grown; specifically, cells are red when grown in green light, when the phycobilisome contains high levels of the red pigmented protein phycoerythrin, and blue-green when grown in red light, when the phycobilisome contains high levels of the blue pigmented protein phycocyanin.

Generating and isolating mutants is a powerful approach for unraveling the biochemical and regulatory processes underlying cellular activities. To identify components of the signal transduction chain that governs CCA, we isolated mutants of the filamentous cyanobacterium *Fremyella diplosphon* (*Calothrix* sp. 7601) that exhibited aberrant CCA. These mutants arise spontaneously, or can be generated by chemical mutagens or as a result of electroporation (Cobley 1983; Tandeau de Marsac 1983; Bruns 1989; Kehoe 1994). Complementation of the mutant strains to

✉ Arthur R. Grossman, Carnegie Institution of Washington, Department of Plant Biology, 260 Panama Street, Stanford, California, 94305, USA (*phone* +01-650-325-1521 (ext. 212); *fax* +01-650-325-6875; *e-mail* arthur@andrew.stanford.edu)
David M. Kehoe, Indiana University, Department of Biology, Bloomington, Indiana, 47401, USA

identify genes responsible for the mutant phenotypes was hindered by difficulties in obtaining high frequency transformation of *F. diplosiphon*. Initially, both conjugation and electroporation were used to transform *F. diplosiphon* (Chiang 1992a; Chiang 1992b), however, the transformation frequencies were very low (25-250 transformants/μg DNA) (Chiang 1992a). Here we present an improved method of transformation using electroporation. The transformation frequencies that we obtain are typically 3×10^3 transformants/μg DNA (Kehoe 1996; Kehoe 1997). This relatively high frequency transformation has enabled us to complement a number of different CCA mutants and define elements required for photoperception and signal transduction.

Materials

Equipment   – Quantum Meter Li-185A with a LI210S quantum sensor (LiCor, Lincoln, Nebraska)
– Gene Pulser Electroporator (Biorad, Hercules, CA)

Procedure

Growth and preparation of cells

1. Inoculate 50 ml of BG-11 buffered with 10 mM HEPES (pH 8.0) (Rippka 1979, except that 0.012 g/l of ferric ammonium citrate was used) with the short filament mutant of *F. diplosiphon*, isolated by John Cobley, in culture tubes that can be aerated.

Note: It is important to use the short filament strain so that the cells grow as individual colonies on plates. The cultures to be transformed should be axenic. Additionally, this and all subsequent steps should be done under sterile conditions since transformation of contaminating bacteria will allow non-transformed cyanobacterial cells to grow on top of the contaminating cells, even on plates containing antibiotics. This results in a high background of non-transformed cyanobacterial cells.

2. Grow the cultures at approximately 20 μmol photons $m^{-2}s^{-1}$ of continuous cool white fluorescent light (measured using a

LiCor Quantum Meter Li-185A containing a LI210S quantum sensor) at 30-32°C and bubbled with a light air stream supplemented with 3-4% CO_2 until the A_{750} is approximately 0.8.

Note: The transformation efficiencies will decrease if the cells are grown beyond this cell density. Also, if the light intensity used exceeds 30-40 µmol photons m^{-2} s^{-1}, polysaccharides will accumulate to high levels on the outer surfaces of the cells, and will not be sufficiently depleted during the dark treatment (see step 3), resulting in low transformation efficiencies.

3. Put the cultures into darkness for 48 h. This is typically achieved by wrapping the culture tubes with aluminum foil. The bubbling of the cultures should continue during this time. The A_{750} does not change significantly during these two days.

Note: Filamentous cyanobacteria often produce extracellular sheaths of polysaccharide, and we postulate that this material physically blocks the passage of DNA into the cells during electroporation. Incubating the cells in darkness for several days is believed to lead to the depletion of the extracellular polysaccharides, making the cell membranes more accessible for interactions with exogenous DNA.

4. Harvest the cells by centrifuging for 10 min at 4-5,000 x g at room temperature. The cells can be collected in a volume of approximately 5 ml by drawing the loose pellet off of the tube bottom with a 10 ml pipette.

5. Transfer the cells to a fresh tube containing 40-45 ml of sterile distilled water (SDW) and resuspend the cells completely. Re-centrifuge and wash the pellet in SDW two more times as in step 4. Because it is filamentous, *F. diplosiphon* does not form a tight pellet. Therefore, after the final wash the cells are contained in approximately 5 ml of SDW. To concentrate the cells further, transfer them to a smaller tube (such as a 15 ml conical polypropylene tube) and re-centrifuge once more as above. Remove the supernatant until a final volume of approximately 1 ml is reached. Resuspend the cells completely in the remaining supernatant and place them on ice. The cells must be used for electroporation within one hour for maximum transformation efficiencies.

Electroporation of cells

1. The plasmids used for transformations are routinely grown in the *E. coli* strain DH5α. The plasmid DNA should be very pure and predominantly (›90%) negatively supercoiled; it is best to use DNA that has been purified on a CsCl gradient (Sambrook 1989) or using a Qiagen plasmid preparation kit (Qiagen, Chatsworth, CA). The DNA should be at a concentration of 2-4 µg/ µl in SDW. Be careful to remove the salt completely from the DNA or it will significantly reduce the electroporation efficiencies. Use 6 µg of DNA per transformation.

Note: We use the shuttle vector pPL2.7 (Chiang 1992a, Cobley 1993) for transforming *F. diplosiphon*. Our insert DNA sizes typically range between 1 and 6 kb. Transformation efficiencies decrease when larger DNA inserts are used.

2. For each electroporation, place 40 µl of cells into a 500 µl microfuge tube and add the appropriate volume of DNA, mixing well. Keep the cells on ice.

3. Place the cells in the cuvette and electroporate each sample, using a Biorad Gene Pulser. The following parameters are most effective: capacitance = 25 µfarads, resistance = 200 ohms, field strength = 10 kilovolts per centimeter of electrode gap. These parameters may need to be varied somewhat depending upon the species being used and the size of the cells. Use Biorad cuvettes with a gap of 1 mm. A new or cleaned/ resterilized cuvette should be used for each transformation.

Note: The time constants obtained under these conditions should be between 4.5 and 5.0 milliseconds. If they are below this, the transformation efficiencies will be reduced significantly.

4. As quickly as possible after each electroporation, place the electroporated solution back into the microfuge tube and place the tubes on ice. After 20 min on ice, transfer the cells to a 50 ml culture tube containing 10 mls of BG-11 and bubble, under conditions previously described, for 8-16 h.

Selection and screening of transformants

1. Harvest the cells after 8-16 hrs by centrifuging for 10 min at 4-5,000 x g. Remove the supernatant, leaving the cells in a few hundred microliters of medium. Resuspend the cells in the remaining medium and plate half of them directly onto BG-11 plates containing the appropriate antibiotic. When transforming with the plasmid pPL2.7, plates containing 25 μg/ml of kanamycin are used. The other half of the cells should be added to 50 ml of BG-11 and bubbled, as described previously, except that the cells should be grown in approximately 40 μmol photons $m^{-2} s^{-1}$.

2. Place the plates at 25-30°C in approximately 40 μmol photons $m^{-2} s^{-1}$ of continuous fluorescent light. Transformants should be visible within two weeks. For the liquid cultures, the appropriate antibiotic should be added incrementally over a six day period. When using pPL2.7, we add kanamycin to 2 μg/ml after 8-16 hours, then to 4 μg/ml (total) after an additional 24 hours, to 10 μg/ml (total) 48 hrs later, and finally, following another 48 hrs of growth, to 25 μg/ml (total).

Note: Selection on plates will allow isolation of independent transformants, while selection in liquid medium will provide transformants more rapidly (approximately one week earlier than on plates).

Troubleshooting

The amount and duration of light received by the cells both before and after electroporation is critical for high frequency transformation. *F. diplosiphon* cells grown in moderate white light (cool white fluorescent bulbs, 60 μmol photons $m^{-2} s^{-1}$) tend to adhere to each other and form large clumps of cells, whereas cells grown in the same light of a lower intensity (20 μmol photons $m^{-2} s^{-1}$) do not aggregate. The amount of light that the cells receive after the electroporation treatment is important as well, because the transformed cells require sufficient energy to grow and survive during the antibiotic selection.

If the time constants for the electroporations are below 4.5 milliseconds, it is usually due to insufficient removal of salt from the DNA. This is not an unusual problem if cesium chloride banding of the DNA is performed and large volumes of the DNA are added to the cell suspension to be electroporated. This is readily solved by reprecipitation of the DNA, followed by thorough washing of the pellet with 70% ethanol.

Comments

We were concerned that a high rate of mutant generation by electroporation could result in the creation of a large number of "false positives" that would make it difficult to identify mutants that were actually complemented. The average frequency of secondary mutant generation after electroporation was determined for several types of CCA mutants and was found to be approximately ten fold below the calculated frequency for complementation of any mutant phenotype with a plasmid library.

Acknowledgements. The authors would like to thank a number of people who have contributed to the development of these methods. In particular, Drs. Brigitte Bruns, Gisela Chiang and Michael Schaefer for establishing the basic electroporation conditions and Drs. Elena Casey, Nicole Tandeau de Marsac, and Jean Houmard for helpful discussions and comments. The development of the protocols presented here was in part supported by a NSF Postdoctoral Fellowship in Plant Biology to DMK and by NSF award MCB 9513576 to ARG. This is Carnegie Institution of Washington publication number 1371.

References

Bruns B, Briggs WR, Grossman AR (1989) Molecular characterization of phycobilisome regulatory mutants of *Fremyella diplosiphon*. Journal of Bacteriology 171:901-908

Cobley JG, Miranda RD (1983) Mutations affecting chromatic adaptation in the cyanobacterium *Fremyella diplosiphon*. Journal of Bacteriology 153:1486-1492

Cobley JG, Zerweck E, Reyes R, Mody A, Seludo-Unson JR, Jaeger H, Weerasuriya S, Navankasattusas S (1993) Construction of shuttle plasmids which can be efficiently mobilized from *Escherichia coli* into the chromatically adapting cyanobacterium, *Fremyella diplosiphon*. Plasmid 30:90-105

Chiang GG, Schaefer MR, Grossman AR (1992a) Transformation of the filamentous cyanobacterium *Fremyella diplosiphon* by conjugation or electroporation. Plant Physiology and Biochemistry 30:315-325

Chiang GG, Schaefer MR, Grossman AR (1992b) Complementation of a red-light-indifferent cyanobacterial mutant. Proc Natl Acad Sci USA 89:9415-9419

Kehoe DM, Grossman AR (1994) Complementary chromatic adaptation: photoperception to gene regulation. Seminars in Cell Biology 5:303-313

Kehoe DM, Grossman AR (1996) Similarity of a chromatic adaptation sensor to phytochrome and ethylene receptors. Science 273:1409-1421

Kehoe DM, Grossman AR (1997) New classes of mutants in complementary chromatic adaptation provide evidence for a novel four-step phosphorelay system. Journal of Bacteriology 179:3914-3921

Rippka R, Deruelles J, Waterbury JB, Herdman M, Stanier RY (1979) Generic assignments, strain histories and properties of pure cultures of cyanobacteria. Journal of General Microbiology 111:1-61

Sambrook J, Fritsch EF, Maniatis T (1989) Molecular cloning: a laboratory manual. second edition. Cold Spring Harbor Laboratory Press, Cold Spring Harbor, New York

Tandeau de Marsac, N (1983) Phycobilisomes and complementary chromatic adaptation in cyanobacteria. Bulletin de L'Institut Pasteur 81:201-254

Abbreviations

A750   absorbance at 750 nm; kb: kilobasepairs; cm: centimeters; mm: millimeters

Electroporation of *Bacillus thuringiensis* and *Bacillus cereus*

JACQUES MAHILLON and DIDIER LERECLUS

Introduction

Contrary to *Bacillus subtilis*, no natural transformation has so far been described for bacteria of the *Bacillus cereus* group (*B. cereus sensu lato*), namely *B. anthracis*, *B. cereus sensu stricto*, *B. thuringiensis* and *B. mycoides*. Moreover, other methods (protoplast, autoplast or vegetative cell transformation) were found to be particularly difficult to apply to these bacteria (Alikhanian et al. 1981, Martin et al. 1981, Fischer et al 1984, Crawford et al. 1987, Heierson et al. 1987).

The first reports of reliable transformation of *B. cereus s.l.* came with the discovery and application of electroporation to bacteria (Neumann et al. 1982, Luchansky et al. 1988): within one year, not less than 5 publications independently reported the successful transformation of *B. thuringiensis* (Bone and Ellar 1989, Lereclus et al. 1989, Mahillon et al. 1989, Masson et al. 1989, Schurter et al. 1989). *B. cereus* (Belliveau and Trevors 1989) and *B. anthracis* (Quinn and Dancer 1990) were also found transformable using similar procedures.

The use of polyethylene glycol (PEG) in preparing bacterial cells for electrotransformation has also been successfully applied to other Gram-positive bacteria, including *Lactobacillus hilgardii* (Josson et al. 1989), *Rhodococcus fascians* (Desomer et al.

✉ Jacques Mahillon, Université catholique de Louvain, Laboratoire de Génétique Microbienne, Place Croix du Sud 2/12, Louvain-la-Neuve, 1348, Belgium (*phone* +32-10-473370; *fax* +32-10-473440; *e-mail* mahillon@mbla.ucl.ac.be)

Didier Lereclus, Institut Pasteur, Unité de Biochimie Microbienne, URA1300 CNRS, 25 rue du Dr. Roux, Paris Cedex 15, 75724, France, Station de Lutte Biologique, INRA, La Minière, Guyancourt, Cedex, 78285, France

1990) or *Streptomyces rimosus* (Pigac and Schrempf 1995). Its beneficial effects have been attributed to volume exclusion, enhancing DNA-cell membrane interactions. It also conveniently plays the role of cryoprotecting agent when conserving the cell at -80°C.

Although optimised for bacteria of the *B. cereus* group, the present PEG-based technique can also be efficiently utilised with other Bacilli such as *B. subtilis* (Mahillon and Kleckner 1992) or *B. brevis* (Okamoto et al. 1997).

Materials

- 100 and 2,000 ml flasks with baffles for bacterial growth
- A thermo-regulated incubation unit with a rotary shaker (New Brunswick)
- Elementary spectrophotometer apparatus for the measurements of cell turbidity (Optical Density at 600 nm)
- A high-voltage electroporation device (Bio-Rad) and the corresponding 2- or 4-mm width cuvettes (Bio-Rad or Eurogentec)

Equipment

- LB growth medium (add 1.5 % Agar for solid media)

Buffers and media

BactoTryptone	1.0 %
Yeast extract	0.5 %
NaCl	1.0 %

- 2XLB growth medium

BactoTryptone	2.0 %
Yeast extract	1.0 %
NaCl	2.0 %

- PEG$_{6000}$ (Freshly prepared in distilled water !)
  Polyethylene glycol (MW 6,000) 40 %
  All concentrations are indicated as % in w/v. All the media are autoclaved for 20 min at 121°C.

Procedure

1. From a freshly grown culture on LB-plate, inoculate 20 ml of liquid 2XLB (or BHI, see the Comment section) with a loop of *Bacillus* cells in a baffle-flask, and incubate overnight at 30°C on a rotary shaker at 180 rpm for aeration.

2. Inoculate 400-ml of liquid LB with 1 ml of the overnight pre-culture in a 2,000-ml flask with baffles.

3. Incubate at 37°C on a rotary shaker at 180 rpm until the culture reaches mid-exponential phase (generally within about 3 to 4 h). This could be monitored by measuring the OD at 600 nm (It generally varies between 0.5 and 1.0, sometime 1.5 depending on the strain used).

4. Centrifuge at 5,000 g the cells at room temperature and re-suspend the pellet into 400 ml of sterile demineralized water. Note that after each centrifugation step, great care should be taken to avoid cell lost when removing the supernatant. In case of problems, re-centrifuge at higher speed and/or longer time. This problem of loose cell pellet is particularly important in the last centrifugation steps.

5. Repeat step 4 twice, recovering the cells successively into 400 and 200 ml of sterile demineralized water.

6. The last pellet is then thoroughly but gently resuspended (do not vortex) into 10 ml of freshly prepared PEG_{6000} 40 % (W/V) in a pre-weighed tube.

7. Centrifuge at 5,000 *g*. Carefully remove all the supernatant and resuspend thoroughly the pellet into V ml of PEG_{6000} 40 %. This volume is determined as follows: V = 1.5 x pellet wet weight (in g). Note that obtaining a perfectly homogenous cell suspension (no cell clump should be left) can be time-consuming and it is best achieved with a P1000 Gilson pipette using a tip whose end has been widened.

8. Aliquot the cell suspension onto multiples of 100 µl. These samples are ready to be used or can be stored as such at -80°C in appropriate freezing tubes.

9. For each transformation experiment, 100 µl of competent cell suspension are mixed with 1 to 5 µl of DNA (between 5 ng to 5 µg) [see comments on the methylation state of the DNA in the Troubleshooting section] in a 2-ml eppendorf tube at room temperature. If the sample is taken from the freezer, it should be left on ice for at least 15 min to allow gentle thawing.

10. Transfer the cell-DNA mixture into a 2-mm width electroporation cuvette avoiding the formation of air bubbles. 4-mm width cuvettes can also be used. The volume of competent cells is then increased to 300-400 µl, and that of the DNA to 10 µl (up to 10 µg).

11. Apply a single electric discharge in a Bio-Rad electroporation apparatus (gene pulser and gene controller) sets with the following parameters:

    - Capacitance: 25 µFa (gene pulser)
    - Resistance: 400 Ω (gene controller)
    - Voltage: 1.4 kV (gene pulser)
    - The time constant should be between 7.0 and 9.5 ms.

Note: When using the 4-mm cuvettes, the electroporation parameters are 25 µFa, 1000 Ω and 2.5 kV.

12. Immediately add 1.9 ml of fresh liquid LB (or BHI) into the cuvette and transfer the mixture using a Pasteur pipette into a 5 ml sterile plastic tube (Falcon) or into a 2.5 ml sterile eppendorf. If the 4-mm cuvettes are utilised, all volumes should be increased accordingly.

13. Incubate the tube at 30 or 37°C for 90 min with gentle shaking.

14. Prepare serial dilution of the cells into LB (generally from 10^0 to 10^{-3}).

15. For the mother tube, spread 50, 100 and 300 µl on LB plates containing the appropriate antibiotics (Table 1). For the dilution tubes, spread 2 x 100 µl on selective plates.

16. Incubate the plates overnight at 30°C or 37°C (depending on growth rate of the bacteria and the specific AB media).

17. The bacterial clones obtained should be transferred to fresh plates to avoid the appearance of background growth. This is particularly true when AB such as Kanamycin or Tetracycline are used (Table 1).

18. Similarly, whenever plasmid transformation is performed in a new strain, its is advisable to confirm its presence in the clone, either by PCR on bacterial colonies (Léonard et al. 1997) or by plasmid minipreparation (Mahillon et al. 1989).

Table 1. Typical AB resistance genes used to transform *B. cereus* and *B. thuringiensis*

Plasmids	Size (kb)	Resistance gene	AB selection	References
pC194	2.9	*cat* from pC194	Cm (5 to 25 µg/ml)	Horinouchi and Weisblum 1982b
pE194	3.7	*erm* from pE194	Ery (5 µg/ml)	Horinouchi and Weisblum 1982a
pHT1618	4.9	*tet* from pBC16	Tet (10 to 20 µg/ml)	Lereclus and Arantes 1992
pHT304	6.5	*ermAM* from Tn*1545*	Ery (5 to 25 µg/ml)	Arantes and Lereclus 1991
pGIC055	6.8	*spc* from Tn*554*	Sp (200 to 400 µg/ml)	Léonard et al 1998
pGIC057	7.1	*aphA3* from *Streptococcus faecalis*	Kan (100 to 150 µg/ml)	Léonard et al 1998

Results

- As is the case for many other bacteria, there is a rather large variation in transformation efficiency depending on the strain and/or plasmid used. However, in most instances, one could expect efficiencies between 10^1 and 10^5 CFU/µg DNA, with an average of 10^3 CFU/µg DNA. It is important to note that the best results are always obtained using freshly prepared cells.

- As indicated below (Troubleshooting section), at least part of the strain-related problems are associated with bacterial restriction of incoming DNA and can be partly solved using non-methylated transforming DNA.

- In our hands, the best strain is *Bacillus thuringiensis* serovar *kurstaki* strain HD73 for which 10^5 - 10^6 CFU/µg DNA can be easily obtained.

- For plasmids carrying large DNA fragments and/or foreign genes, transformation frequency can be drastically reduced. In these cases, the necessity for large amounts of DNA required the use of 4-mm cuvettes.

- Obviously, the transformation efficiency depends on the purity of the plasmid DNA preparation. However, most current *Escherichia coli* or *Bacillus* minipreparation protocols will give suitable plasmid DNA. Highly purified DNA (i.e. CsCl-EtBr gradient grade) will be sought only in critical situations.

- Due to the rather low efficiency generally observed in *B. cereus/B. thuringiensis* transformation, its application is so far mostly restricted to plasmid transformation. Experiments dealing with either homologous or transpositional chromosome or plasmids integration cannot be obtained in a single step. However, several thermosensitive gram-positive plasmids have recently been developed, enabling gene replacement or transpositional insertion (see for instance: Léonard et al. 1998, Lereclus et al. 1992, Steinmetz and Richter 1994).

Troubleshooting

- In the preculture step, it is important to avoid the onset of sporulation before inoculating the mother culture. This can be achieved by using either a 2-fold concentrated LB (referred to as 2XLB) or BHI (Brain Heart Infusion - Difco).

- As indicated above, the wide variation in transformation efficiency among *B. thuringiensis* and *B. cereus* strains is thought to be due, in part at least, to restriction of incoming DNA by the cell enzymatic machinery. Macaluso and Mettus (1991) have indeed demonstrated that DNA isolated from either *B. thuringiensis*, *Bacillus megaterium* or a Dam⁻/

Dcm⁻ *E. coli* strain significantly improved *B. thuringiensis* transformation efficiency (up to two orders of magnitude). Consequently, whenever some difficulties are foreseen, plasmid DNA should be first prepared from a Dam⁻/Dcm⁻ *E. coli* strain.

- Never use DNA in salty solutions: this can cause arcing during electroporation. If a DNA ligation mixture is used, it is always better to either desalt the preparation on a dialysis membrane or a 0.05 µm Millipore filter [20 µl are deposited on the membrane/filter floating on a water solution for 15 min] or precipitate the DNA. If the volume is too small, use no more than 1µl of DNA ligation solution for electroporation.

- It is particularly important to use freshly prepared PEG solution (less than one-month old).

- If the number of CFU is extremely low, it is always possible to concentrate the suspension by centrifuging the cells before spreading and to recover them in 200 µl of LB.

Comments

Note that contrary to what is used for many other bacteria, all the steps of the *Bacillus* protocol are performed at room temperature (except for the cell growth). Working on ice does not generally improve the efficiency.

Although the PEG_{6000} gives excellent results, other PEG will give good results as well. In general, PEG_{1000} to PEG_{20000} at concentrations varying between 30 and 40 % could be tested. For some strains, this might even be a parameter to be changed for optimising the results.

The electroporation parameters (resistance, capacitance or voltage) can also be modified, but at least in our hands, few improvements have so far been obtained.

Although small (< 15 kb) molecules are generally used for cell transformation, plasmids of more than 150-kb have also been reported to occur at comparable frequency (Belliveau and Trevors 1989).

Acknowledgements. We are very grateful to C. Léonard, Y. Chen and V. Sanchis for their useful comments and suggestions on the manuscript. JM is Research Associate at the FNRS (Fonds National pour la Recherche Scientifique), Belgium and DL is Scientist at INRA (Institut National de Recherche Agronomique), France.

References

Alikhanian SI, Ryabchenko NF, Bukanov NO, Sakanyan VA (1981) Transformation of *Bacillus thuringiensis* subsp. *galleriae* protoplasts by plasmid pBC16. J Bacteriol 146:7-9

Arantes O, Lereclus D (1991) Construction of cloning vectors for *Bacillus thuringiensis*. Gene 108:115-119

Belliveau BH, Trevors JT (1989) Transformation of *Bacillus cereus* vegetative cells by electroporation. Appl Environ Microbiol 55:1649-1652

Bone EJ, Ellar DJ (1989) Transformation of *Bacillus thuringiensis* by electroporation. FEMS Microbiol Lett 49:171-177

Crawford IT, Greis KD, Parks L, Streips UN (1987) Facile autoplast generation and transformation in *Bacillus thuringiensis* subsp. *kurstaki*. J Bacteriol 169:5423-5428

Desomer J, Dhaese P, Van Montagu M (1990) Transformation of *Rhodococcus fascians* by high-voltage electroporation and development of *R. fascians* cloning vectors. Appl Environ Microbiol 56:2818-2825

Fischer HM, Lüthy P, Schweitzer S (1984) Introduction of plasmid pC194 into *Bacillus thuringiensis* by protoplast transformation and plasmid transfer. Arch Microbiol 139:213-217

Heierson A, Landen R, Lövgren A, Dalhammar G, Boman HG (1987) Transformation of vegetative cells of *Bacillus thuringiensis* by plasmid DNA. J Bacteriol 169:1147-1152

Horinouchi S, Weisblum B (1982a) Nucleotide sequence and functional map of pE194, a plasmid that specifies inducible resistance to macrolide, lincosamide, and streptogramin type B antibiotics. J Bacteriol 150:804-814

Horinouchi S, Weisblum B (1982b) Nucleotide sequence and functional map of pC194, a plasmid that specifies inducible chloramphenicol resistance. J Bacteriol 150:815-825

Josson K, Scheirlinck T, Michiels F, Platteeuw C, Stanssens P, Joos H, Dhaese P, Zabeau M, Mahillon J (1989) Characterization of a gram-positive broad-host-range plasmid isolated from *Lactobacillus hilgardii*. Plasmid 21:9-20

Léonard C, Chen Y, Mahillon J (1997) Diversity and differential distribution of IS*231*, IS*232* and IS*240* among *Bacillus cereus*, *B. thuringiensis*, and *B. mycoides*. Microbiology 143:2537-2547

Léonard C, Zekri O, Mahillon J (1998) Integrated physical and genetic mapping of *Bacillus cereus* and other Gram$^+$ bacteria based on IS*231*A transposition vectors. Infect Immun *In press*.

Lereclus D, Arantes O, Chaufaux J, Lecadet M (1989) Transformation and expression of a cloned delta-endotoxin gene in *Bacillus thuringiensis*. FEMS Microbiol Lett 51:211-217

Lereclus D, Arantes O (1992) *spbA* locus ensures the segregational stability of pTH1030, a novel type of gram-positive replicon. Mol Microbiol 6:35-46

Lereclus D, Vallade M, Chaufaux J, Arantes O, Rambaud S (1992) Expansion of insecticidal host range of *Bacillus thuringiensis* by *in vivo* genetic recombination. Biotechnology 10:418-421

Luchansky JB, Muriana PM, Klaenhammer TR (1988) Application of electroporation for transfer of plasmid DNA to *Lactobacillus, Lactococcus, Leuconostoc, Listeria, Pediococcus, Bacillus, Staphylococcus, Enterococcus* and *Propionibacterium*. Mol Microbiol 2:637-646

Macaluso A, Mettus AM (1991) Efficient transformation of *Bacillus thuringiensis* requires nonmethylated plasmid DNA. J Bacteriol 173:1353-1356

Mahillon J, Chungjatupornchai W, Decock J, Dierickx S, Michiels F, Peferoen M, Joos H (1989) Transformation of *Bacillus thuringiensis* by electroporation. FEMS Microbiol Lett 60:205-210

Mahillon J, Kleckner N (1992) New IS10 transposition vectors based on a gram-positive replication origin. Gene 116:69-74

Martin PA, Lohr JR, Dean DH (1981) Transformation of *Bacillus thuringiensis* protoplasts by plasmid deoxyribonucleic acid. J Bacteriol 145:980-983

Masson L, Préfontaine G, Brousseau R (1989) Transformation of *Bacillus thuringiensis* vegetative cells by electroporation. FEMS Microbiol Lett 51:273-277

McDonald IR, Riley PW, Sharp RJ, McCarthy AJ (1995) Factors affecting the electroporation of *Bacillus subtilis*. Appl Bacteriol 79:213-218

Neumann E, Schaefer-Ridder M, Wang Y, Hofschneider PH (1982) Gene transfer into mouse lyoma cells by electroporation in high electric fields. EMBO J 1:841-845

Okamoto A, Kosugi A, Koizumi Y, Yanagida F, Udaka S (1997) High efficiency transformation of *Bacillus brevis* by electroporation. Biosci Biotechnol Biochem 61:202-203

Pigac J, Schrempf H (1995) A simple and rapid method of transformation of *Streptomyces rimosus* R6 and other Streptomycetes by electroporation. Appl Environ Microbiol 61:352-356

Quinn CP, Dancer BN (1990) Transformation of vegetative cells of *Bacillus anthracis* with plasmid DNA. J Gen Microbiol 136:1211-1215

Scheirlinck T, Mahillon J, Joos H, Dhaese P, Michiels F (1989) Integration and expression of alpha-amylase and endoglucanase genes in the *Lactobacillus plantarum* chromosome. Appl Environ Microbiol 55:2130-2137

Schurter W, Geiser M, Mathe D (1989) Efficient transformation of *Bacillus thuringiensis* and *B. cereus* via electroporation: transformation of acrystalliferous strains with a cloned delta-endotoxin gene. Mol Gen Genet 218:177-181

Steinmetz M, Richter R (1994) Easy cloning of mini-Tn10 insertions from the *Bacillus subtilis* chromosome. J Bacteriol 176:1761-1763

Suppliers

Bio-Rad Laboratories
2000 Alfred Nobel Drive
Hercules, CA 94547
USA
phone: +1-510-7411000
fax: +1-510-7415800
URL: www.bio-rad.com

Difco Laboratories
PO Box 331058
Detroit, MI 48232-7058
USA
phone: +1-313-4628500
fax: +1-313-4628517

Eurogentec Bel s.a.
Parc Scientifique de Sart Tilman
B-4102 Seraing
Belgium
phone : +32-4-3660150
fax : +32-4-3655103
email: order@eurogentec.be

New Brunswick Scientific Inc
44 Talmadge Road
Edison, NJ 08818-4005
USA
phone: +1-908-2871200
fax: +1-908-2874222

UCB-VEL s.a.
Geldenaaksebaan 464
B-3030 Leuven
Belgium
phone : +32-16-281811
fax : +32-16-281861

▨ Abbreviations

AB	antibiotic(s)
CsCl-EtBr	Caesium Chloride-Ethidium Bromide
CFU	Colony Forming Unit
Ery	Erythromycin
Fa	Faraday
g	gravity acceleration constant
Kan	Kanamycin
kV	kilovolt
mm	millimetre
ms	millisecond
MW	Molecular Weight
OD	Optical density, measured at 600 nm
PEG$_{6000}$	Polyethylene glycol with an average MW of 6,000
s.l.	*sensu lato*
Sp	Spectinomycin
s.s.	*sensu stricto*
Tet	Tetracycline
w/v	weight/volume

Introduction of Plasmids into *Azospirillum brasilense* by Electroporation

ANN VANDE BROEK and JOS VANDERLEYDEN

Introduction

The genus *Azospirillum* comprises Gram⁻ N_2 fixing soil bacteria living in close association with the roots of numerous plants. Field trials, carried out at different locations have demonstrated significant plant growth promotion upon *Azospirillum* inoculation. Based on 16S rRNA similarity, the genus *Azospirillum* has been classified within the alpha subdivision of Proteobacteria. Plasmids of the IncP group are stably maintained in a wide range of *Azospirillum* strains and can be succesfully transferred to *Azospirillum* either by conjugation or by electroporation. As compared to conjugation, the electroporation technique offers the advantages of being much faster and avoiding problems with counterselection of the *Escherichia coli* donor strain.

Materials

- Gene Pulser electroporation apparatus (Bio-Rad Laboratories, Richmond, CA, USA)   **Equipment**
- Electroporation cuvettes (0.2 cm interelectrode gap, e.g. Eurogentec (Seraing, Belgium; Abingdon, United Kingdom))
- A plasmid preparation kit allowing isolation of highly pure plasmid DNA (e.g. S.NA.P. miniprep kit from Invitrogen (Leek, The Netherlands; Carlsbad, CA, USA))

Ann Vande Broek, F.A. Janssens Laboratory of Genetics, Kardinaal Mercierlaan 92, Heverlee, 3001, Belgium
✉ Jos Vanderleyden, F.A. Janssens Laboratory of Genetics, Kardinaal Mercierlaan 92, Heverlee, 3001, Belgium (*phone* +32-16321631; *fax* +32-16321966; *e-mail* Jozef.Vanderleyden@agr.kuleuven.ac.be)

Media **YEP broth**
- 10 g/l Bacto peptone
- 5 g/l NaCl
- 10 g/l Yeast extract

Procedure

Preparation of cells

1. Grow cells at 30°C with shaking to mid-exponential growth phase (OD600 = 0.6 to 0.8) in 1 liter of YEP broth

2. Chill flask on ice and centrifuge cells in a cold rotor at 6000 rpm for 10 minutes.

3. Resuspend pellet in 1 volume of cold MilliQ water. Centrifuge as above.

4. Resuspend pellet in 1/2 volume of cold MilliQ water. Centrifuge as above.

5. Resuspend pellet in 1/20 volume 10 % glycerol. Centrifuge as above.

6. Resuspend cells in a final volume of 1/50 to 1/200 volume of 10 % glycerol (final cell concentration of 10^9 to 5×10^{10} cells/ml)

Note: A decrease in transformation efficiency is observed when cell concentrations higher than 5×10^{10} cells/ml are used.

7. Store 0.2 ml aliquots at -80°C.

Electrotransformation

1. Set the Gene Pulser apparatus to the 25 µF capacitor and 1.5 to 2 kV and set the Pulse Controller to 400 Ω (optimal for the *A. brasilense* strain 7030).

2. Chill 0.2 cm electroporation cuvette on ice.

3. Thaw cells on ice.

4. Mix 40 µl of cell suspension with 50 to 150 ng of plasmid DNA while on ice.

Note: It is essential to use highly pure plasmid DNA!

5. Transfer mixture of cells and DNA to the bottom of the ice-cold electroporation cuvette.

6. Apply one pulse. This should result in a pulse with a time constant of 9 to 9.5.

7. Immediately resuspend in 1 ml of YEP medium, transfer to a tube and incubate with shaking at 30 °C for 5 hours.

8. Plate on selective medium and incubate at 30 °C for 2 to 3 days.

Table 1. Effect of pulse length and electric field strength on the electroporation-induced transformation of *A. brasilense* 7030 with pRK290 plasmid DNA (Ditta et al., 1980)

Capacitance (µF)	Pulse Controller resistance (Ω)	Field strength (kV/cm)	Efficiency[a]	Survival[b]
25	400	5	1.4×10^3	ND[c]
25	400	6	3.4×10^3	ND
25	400	7.5	1.5×10^4	± 10 %
25	400	8.5	2.2×10^4	ND
25	400	10	5.7×10^3	ND
25	400	11	10^2	ND
25	400	12.5	2.5×10^2	ND
25	200	10	5.1×10^2	± 5 %
25	200	12.5	2.8×10^3	± 5 %

[a] Calculated as number of transformants per µg DNA
[b] Calculated as (viable cells after shock/total input cells)X100
[c] ND, not determined

Results

Using the optimized protocol described here, transformation efficiencies of 10^3 to 10^4 transformants/ µg DNA are usually obtained with the *Azospirillum brasilense* strain 7030 and with IncP plasmids of 20 kb (RP4 derivatives: pRK290 (Ditta et al., 1980) and pLAFR3 (Staskawicz et al., 1987)). Table I shows typical efficiencies of 7030 electroporation with pRK290 plasmid DNA as a function of the applied field strength and pulse length.

The purity of the plasmid DNA was observed to be critical for efficient electroporation. In our hands, a preparation kit yielding highly pure plasmid DNA (such as e.g. the S.N.A.P. miniprep kit from Invitrogen) always gave higher transformation efficiencies as compared to plasmid DNA isolated according to the alkaline lysis method (Sambrook et al., 1989). Using the above described protocol, *A. brasilense* 7030 was found to be tranformed with plasmids up to 25 kb. Several attempts, however, to electroporate this strain with an IncP plasmid of 40 kb have so far been unsuccessful.

Acknowledgements. A.V.B. expresses her gratitude towards the Flemish 'Fonds voor Wetenschappelijk Onderzoek' for her postdoctoral fellowship.

References

Ditta G, Stanfield S, Corbin D and Helinski DR (1980) Broad host range DNA cloning system for gram-negative bacteria: construction of a gene bank of *Rhizobium meliloti*. Proc. Nat. Acad. Sci. USA 77: 7347-7351
Sambrook J, Fritsch EF and Maniatis T (1989) Molecular cloning: a laboratory manual. Cold Spring Harbor, New York
Staskawicz B, Dahlbeck D, Keen N and Napoli C (1987) Molecular characterization of cloned avirulence genes from race 0 and race 1 of *Pseudomonas syringae* pv. *glycinea*. J. Bacteriol. 169: 5789-5794.

Cyanobacteria: Electrotransformation and Electroextraction

TOIVO KALLAS

Introduction

Cyanobacteria are an ancient, diverse, and ecologically important group of oxygenic-photosynthetic eubacteria related phylo-
genetically to plant chloroplasts (Rippka et al. 1979; Wilmotte
1994). Some members of unicellular genera such as *Synechococcus* and *Synechocystis* have physiological mechanisms for DNA
uptake and are readily transformed genetically (Porter 1986;
Shestakov and Reaston, 1987). DNA can be introduced into additional strains, including filamentous forms capable of nitro-
gen-fixation and heterocyst differentiation, by conjugal transfer
from *Escherichia coli* mediated by broad host-range conjugative
plasmids (Elhai and Wolk, 1988). Many other cyanobacteria with
interesting properties appear incapable of biological DNA uptake. For these, electroporation offers an approach that has
widely allowed the transformation of diverse bacteria (Halloway
1993). Even where other methods are possible, electroporation
may be advantageous because of simplicity or efficiency. Electrotransformation of cyanobacteria was first demonstrated by
Thiel and Poo (1989) and has since been applied to several strains
(Chiang et al. 1992; Moser et al. 1993; Mühlenhoff and Chauvat
1996; Chapter 29). Electroporation in addition provides an efficient method for extracting macromolecules from cells.

This chapter describes two protocols that work effectively for
Nostoc sp. PCC 7121, a largely unicellular mutant of *Nostoc* sp.
PCC 7906 (Moser et al. 1993). The electrotransformation proce-
dure is based on Thiel and Poo (1989). The "electroextractio"

Toivo Kallas, University of Wisconsin-Oshkosh, Department of Biology
and Microbiology, Oshkosh, WI, 54901, USA (*phone* +01-920-424-7084;
fax +01-920-424-1101; *e-mail* kallas@uwosh.edu)

protocol allows rapid release of plasmid and chromosomal DNA from cells for plasmid isolation or DNA amplification by polymerase chain reaction (PCR). These procedures should be widely adaptable to other cyanobacteria. Guidelines are discussed for electrotransformation of previously untransformed cyanobacteria and for optimization.

Outline

The procedure for electrotransformation of *Nostoc* PCC 7121 is outlined in Figure 1.

Materials

Equipment
- Electroporator capable of a variable output voltage and pulse duration. (e.g. "Gene Pulser" with "Pulse-Controller" [variable resistor], Bio-Rad Laboratories, Hercules, CA 94547, U.S.A., www.bio-rad.com)
- Electroporation cuvettes (0.2 or 0.1 cm gap width). (e.g. Equi-Bio ECU-102, Kent, UK ME17 4LT, e-mail: 101552.300@compuserve.com)
- Constant temperature heating block
- Constant temperature incubator fitted with cool-white fluorescent lamps.

Note: Many cyanobacteria grow well at ambient temperature without the need for incubators.

- Spectrophotometer or Colorimeter (e.g. Klett-Summerson with no. 66 red filter)
- Centrifugal cuvette drier (e.g. Roto-Vette model 310, Hellma GmbH, Mühllheim, Germany, P.O. Box 1163)

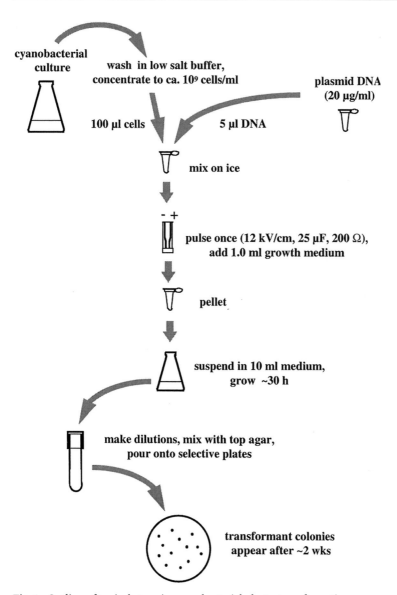

cyanobacterial
culture

wash in low salt buffer,
concentrate to ca. 10⁹ cells/ml

plasmid DNA
(20 μg/ml)

100 μl cells   5 μl DNA

mix on ice

- +

pulse once (12 kV/cm, 25 μF, 200 Ω),
add 1.0 ml growth medium

pellet

suspend in 10 ml medium,
grow ~30 h

make dilutions, mix with top agar,
pour onto selective plates

transformant colonies
appear after ~2 wks

Fig. 1. Outline of typical steps in cyanobacterial electrotransformation.

Media and buffers – BG-11 (ATCC 616) medium[a]

NaNO3	1.5 g
K2HPO4	0.04 g
MgSO4 7H2O	0.0075 g
CaCl2 2H2O	0.036 g
Citric acid	0.006 g
Ferric ammonium citrate	0.006 g
EDTA (disodium salt)	0.001 g
NaCO3	0.02 g
Trace metal mix A5	1.0 ml
Double distilled water	1.0 L

[a] From Rippka et al. (1979). The pH should be 7.1 after sterilization. Recipes for other media may be found at Cyanosite (http://www-cyanosite.bio.purdue.edu).

– Trace metal mix A5

H3BO3	2.86 g
MnCl2 4H2O	1.81 g
ZnSO4 7H2O	0.222 g
NaMoO4 2H2O	0.39 g
CuSO4 5H2O	0.079 g
Co(NO3)2 6H2O	49.4 mg
Double distilled water	1.0 L

– BG-11 agar plates[a]
  – Make 2X BG-11 medium in 0.5 L double distilled water (ddH$_2$O).
  – In a 2 L flask containing a stir bar, make 2X agar (12 g agar in 0.5 L ddH$_2$O)
  – Autoclave 20 min at 120 °C and cool to 50 °C.
  – Add the 2X BG-11 to the agar flask, add antibiotic as needed, stir to mix, and pour into Petri plates.

- [a]Separate autoclaving of mineral salts and agar greatly improves the plating efficiency of many cyanobacteria (Allen et al. 1968)
- Antibiotic stock: Neomycin (Nm), 50 mg/ml in ddH$_2$O, is filter sterilized (0.22 μm pore) and stored at -20 °C. See Sambrook et al. (1989) for other antibiotic stocks.
- 2X top agar
  - Make a 2% mixture of bacteriological agar (e.g. Difco) in double-distilled water.
  - Autoclave 20 min at 120 °C.
  - Cool to 50 °C and dispense sterile, 2 ml portions into sterile, capped, glass culture tubes. (e.g. 13 mm dia). Store at 4 °C.
  - Before use, boil 5 min in a water bath to melt the agar. Equilibrate to 47 °C by holding the tubes in a 47 °C heating block at least 15 min before use.
- Luria Bertani (LB) medium[a]

Bacto-tryptone	10 g
Bacto-yeast extract	5 g
NaCl	10 g
ddH$_2$O to 1.0 L, adjust pH to 7.0 with NaOH (ca. 0.2 ml 0.5 N)	

[a] (Sambrook et al. 1989)

- S.O.C. medium[a]

Bacto-tryptone	20 g
Bacto-yeast extract	5 g
NaCl	0.5 g

Dissolve the above in 950 ml ddH$_2$O.

Add 10 ml 250 mM KCl and adjust pH to 7.0 with NaOH (ca. 0.2 ml 0.5 N).

Adjust the volume to 1.0 L and autoclave.

After cooling, add 5 ml sterile 0.2 M MgCl$_2$ and 20 ml filter-sterilized 1.0 M glucose.

[a] (Sambrook et al. 1989)

▦ Procedure

Electrotransformation of cyanobacteria

Preparation of electrocompetent cells

1. Inoculate 50 ml of medium (125 ml flask) and grow cells to late log phase.

Note: *Nostoc* PCC 7121 is grown in BG-11 medium with gentle agitation or bubbling at 28 °C under cool-white fluorescent lamps (500-1000 lx). A flow of 1% CO_2 in air increases the growth rate but is optional. For *Nostoc,* late log phase corresponds to 100-200 Klett Units [ca. 1.0-2.0 O.D.$_{750nm}$, 1 Klett Unit = ca. 4×10^4 cells/ml.] The preparation may be scaled up or down as needed.

2. Chill cells on ice. Pellet cells by centrifugation (6000 x g, 4 °C, 5 min). Discard supernatant.

3. Suspend pellet in 50 ml ice-cold 1 mM HEPES pH 7.3 buffer.

4. Wash twice and suspend the final pellet to ca. 10^9 cells/ml in the same buffer.

Note: Cells from a 50 ml *Nostoc* culture at 200 Klett Units would be suspended in 0.5 ml. During washing steps the pellet becomes loose. Final concentration is best done in a microcentrifuge. Electrocompetent *Nostoc* may be·suspended in buffer/10% glycerol and stored at -80° C but the transformation efficiency is ca. 1/10 that of fresh cells.

Electroporation

1. Cool the cuvette holder on ice and have ready all cuvettes, tubes, and flasks.

Note: Cell recovery is optimized by use of ice-cold cuvettes, cell suspensions, and DNA solutions, and by rapid dilution of cells into growth medium after electroporation.

2. Adjust the electroporator to settings appropriate for the strain.

Note: Settings are determined empirically. For *Nostoc* PCC 7121, we use 8 - 12 kV/cm (1600 - 2400 V in an 0.2 cm gap width cuvette), 25 μF, and 200 Ω. This generates a pulse time constant of ca. 4.5 ms in the Bio-Rad Gene Pulser.

3. Add 1-5 μl plasmid DNA to 100 μl of electrocompetent cells on ice. Mix with pipet tip.

Note: Avoid making bubbles. The DNA concentration should be at least 5 µg/ml and suspended in water or a low salt buffer such as 10 mM Tris-HCl pH 7.5, 0.1 mM EDTA.

4. Transfer the cell/DNA mixture to a cold, 0.2 cm gap electroporation cuvette on ice and incubate for about 1 min.

5. Place the cuvette into the electroporation chamber. Pulse once.

6. Immediately add 1 ml sterile growth medium to the cuvette. Use a Pasteur pipet to suspend the cells and transfer into a microcentrifuge tube.

Note: To ensure efficient cell removal, 0.5 ml medium may be added first and the cuvette rinsed with an additional 0.5 ml medium. Electroporation cuvettes may be reused several times if they are labeled to avoid cross-contamination, rinsed with sterile ddH$_2$O and ethanol after electroporation, and filled with ethanol during storage. Cuvettes are conveniently dried prior to use by centrifugation in a Hellma Roto-Vette cuvette drier.

1. Centrifuge the electroporated cells 14,000 x g for 1 min, suspend in 10 ml growth medium in a 25 ml flask, and incubate with gentle agitation under growth conditions.

Cell recovery and plating

Note: This incubation allows expression of antibiotic-resistance or other necessary genes and should correspond to approximately 1-3 doubling times of a standard culture. *Nostoc* PCC 7121 is incubated for 30 h.

2. Pellet cells by centrifugation, resuspend in fresh medium, and prepare dilutions in 2ml volumes of growth medium.

Note: For *Nostoc* PCC 7121, transformation frequencies may be as high as several per 10^3 cells and 2 ml of a 10^{-2} dilution may yield hundreds of colonies per plate.

3. Have ready tubes containing 2 ml molten 2% top agar in a heating block at 47 °C. Add antibiotics (as needed) to appropriate concentrations.

4. Add 2 ml of electroporated cell culture (at an appropriate dilution) to a tube containing 2 ml top agar, mix gently by vortex, and pour onto selective agar plates.

5. Allow top agar to solidify 5 min. Invert plates and incubate under growth conditions.

Note: For *Nostoc* PCC 7121 transformed with plasmid pRL25, BG-11 agar plates containing 12 µg Nm/ml are used for selection. Transformant colonies appear after about two weeks. High light may cause photo-damage and transformation plates are typically incubated for the first 1-2 da at lower light intensity.

6. When colonies reach 1-2 mm diameter, they may be transferred to liquid medium.

Note: *Nostoc* pRL25 transformants are first transferred to medium containing 4 µg Nm/ml which is raised to 12-50 µg/ml after the culture reaches a density of e.g. 50 Klett Units. The success of transfer to liquid is often improved if colonies are transferred together with a small surrounding block of agar. (Samples of liquid cultures may be stored frozen at -80 °C by addition of 10% glycerol or 7% DMSO.)

Electroextraction of plasmid and genomic DNA

Electroextraction

1. Pellet (1-2 min in a microcentrifuge) 1.5 ml of a late-log or stationary phase culture.

Note: A colony from an agar plate may be suspended in sterile ddH$_2$O and used as the starting material.

2. Resuspend cell pellet in 1.0 ml ice-cold 1.0 mM HEPES pH 7.3 buffer.

3. Repeat the washing step twice and resuspend in a total volume of 40-60 µl of the same buffer.

4. Transfer to a chilled 0.1 cm gap width electroporation cuvette and pulse once at 24 kV/cm (25 µF and 200 Ω).

Plasmid DNA recovery

1. Add 40 µl (ca. 10^9 cells) of appropriate, ice-cold *E. coli* electrocompetent cells to the cyanobacterial electroextract from step 4 above.

2. Mix by pipet tip and pulse once at 10 kV/cm (25 µF and 200 Ω).

3. Immediately add 1.0 ml sterile S.O.C. medium, transfer to a culture tube and incubate with shaking 1 h at 37 °C.

4. Spread the entire sample onto a selective LB plate, allow the liquid to dry and incubate at 37 °C. Colonies appear after overnight incubation.

Note: Cyanobacterial/*E. coli* shuttle plasmids are recovered from *E. coli* by standard alkaline lysis (Sambrook et al. 1989) or commercial "minipreparations" such as the "Wizard™" procedure of Promega (Madison, WI, USA).

1. Transfer the cyanobacterial "electroextract" from step 4 above into a microcentrifuge tube and centrifuge 1-2 min. **PCR amplification**

2. Use 5-20 µl of the supernatant in a PCR reaction.

Note: PCR amplification of electroextracted *Nostoc* genomic DNA may be performed as described in Moser et al. (1995). We have recently obtained excellent results with the "simple hot-start" protocol and PCR reagents of Qiagen GmbH (40724 Hilden, Germany).

Results

Electrotransformation

Electroporation as described above provides a mechanism for the physical entry of DNA into *Nostoc* sp. PCC 7121 and probably most other cyanobacteria. Whether the introduced DNA will lead to genetic transformation (i.e. to expression of antibiotic resistance or other phenotypes) depends on additional factors such as host nucleases, stable replication or integration of the introduced DNA, and adequate expression of genetic markers (Elhai and Wolk 1988; Moser et al. 1993; Thiel 1994). Table 1 shows the results of typical electrotransformations of *Nostoc* sp. PCC 7121 and illustrates the importance of protection against host endonucleases. Transformation to Nm-resistance is improved by approximately 100-fold when the transforming plasmid pRL25 (Wolk et al. 1988) is copurified from *E. coli* carrying both pRL25 and a second plasmid, pRL528 (Elhai and Wolk, 1988). The latter encodes *Eco*47II and *Ava*I methylases that protect DNA from cleavage by the *Ava*II and *Ava*I endonucleases of many *Anabaena* and *Nostoc* strains. *Nostoc* PCC 7121 carries *Nsp*7121I which is an isoschizomer of *Ava*II and *Eco*47II (Moser et al. 1993). Other factors essential for stable transformation are discussed below.

Table 1. Electrotransformation of *Nostoc* sp. PCC 7121[a]

Transforming DNA	Field strength (kV/cm)	No. cells plated	No. NmR colonies	Transformation frequency
pRL25	9.0	10^6	4	4×10^{-6}
pRL25/ pRL528	9.0	10^5	150	2×10^{-3}
pRL25/ pRL528	0	10^8	0	0
none	9.0	10^8	0	0

[a] Electroporated Nostoc cells were plated in triplicate. Transformation frequencies are expressed as no. Nm^R colonies per viable cell. Adapted from Moser et al. (1993).

Electroextraction

Figure 2 shows absorption spectra of the suspending medium after high-voltage electroporation of *Nostoc* PCC 7121. Above a threshold voltage, increasing quantities of nucleic acids and proteins such as the brilliantly colored, water-soluble, light-harvesting phycobiliproteins (Sidler 1994) are released as a function of applied voltage. Shuttle plasmids such as pRL25 are easily recovered from such electroextracts by subsequent electroporation of *E. coli* and the released DNAs serve well as templates for PCR amplification (Moser et al. 1995).

Electrotransformation of "novel" cyanobacteria

Phycobiliprotein release in addition provides a visual cue for conditions that may allow electroporation of previously untransformed strains. For example, *Nostoc* PCC 7121 is optimally electroporated at voltages of 10-12 kV/cm which begin to release

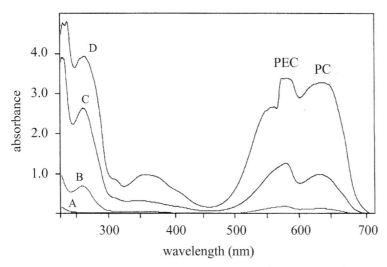

Fig. 2. Absorption spectra of material released after electoporation of *Nostoc* sp. PCC 7121. Traces A-D were obtained from undiluted supernatants of suspensions electroporated at 0, 6, 12, and 18 kV/cm, respectively. PEC and PC mark the regions of the absorption maxima of the phycoerythrocyanin and phycocyanin pigments of *Anabaena* and *Nostoc* strains (Sidler, 1994). Adapted from Moser et al. (1995).

substantial amounts of phycobiliproteins from these cells (trace B, Fig. 2). Some cyanobacteria may be resistant to electroporation. We have been unable to improve by electroporation the transformation of *Synechococcus* sp. PCC 7002. Notably, this strain did not release detectable phycobiliproteins even at 24 kV/cm and the transformants obtained may have resulted entirely from the native, physiological DNA uptake mechanism (Porter 1986). It may be possible to render such tough cyanobacteria "electrocompetent" by pretreatments with lysozyme or mild sonication. The latter has been used to generate electrocompetent cells of the filamentous bacterium, *Saccharopolyspora erythraea* (Fitzgerald et al. 1998). Once some electro-transformants are obtained, conditions may be optimized for the field strength and pulse duration that give the highest number of transformants per viable cell (Thiel and Poo 1989).

Considerations for establishing gene transfer into new strains have been reviewed by Thiel (1994). Briefly, in addition to DNA entry, one needs: i) "transforming" DNA (typically a plasmid) capable of autonomous replication or integration into the host genome, ii) a genetic marker on the transforming DNA that can be expressed and detected in the host, and iii) typically a way to circumvent host restriction or other nucleases. As discussed above, electroporation should permit DNA entry into most cyanobacteria. The extensive polysaccharide capsule of some cyanobacteria may entrap DNA and will probably need to be removed. For *Microcystis* spp. this can be accomplished by washes in sterile distilled water (Plude et al. 1991). Electrotransformation trials might begin with broad host-range plasmids such as RK2 or RSF1010 or their derivatives that replicate and express antibiotic-resistance in many cyanobacteria (Elhai and Wolk, 1988; Mühlenhoff and Chauvat, 1996). Alternatively, integration vectors may be constructed by cloning cyanobacterial DNA fragments into *E. coli* plasmids (Mühlenhoff and Chauvat, 1996). Protection against nucleases may be achieved by premethylating transforming DNA via S-adenosyl methionine and crude cyanobacterial extracts (Swanson et al. 1992) or simply by saturating nucleases with large amounts of donor or non-specific carrier DNAs. The latter has been used to obtain efficient electrotransformation of the nuclear genome of the alga, *Chlamydomonas reinhardtii* (Shimogawara et al. 1998). A nice recent example of gene transfer into a novel cyanobacterium is the electrotransformation with RSF1010-derived and integrative plasmids of the thermophile, *Synechococcus elongatus* (Mühlenhoff and Chauvat, 1996).

Finally, it should be noted that electroporation is mutagenic in at least some cyanobacteria (Bruns et al. 1989). Mutagenesis appears to result from induction of endogenous transposable elements (Kahn et al. 1997). This may be problematic but also provides a potential tool for the isolation of pigment and other mutants (Bruns et al. 1989; Kahn et al. 1997).

Acknowledgements. Much of the electroporation work in the author's lab was performed by Duane Moser and Daniel Zarka and supported by NSF grant RUI DMB-8902695.

References

Allen M (1968) Simple conditions for growth of unicellular blue-green algae on plates. J Phycol 4: 1-4

Chiang G, Schaefer M, Grossman A (1992) Transformation of the Filamentous Cyanobacterium Fremyella diplosiphon by Conjugation or Electroporation. Plant Physiol Biochem 30: 315-325

Elhai J, Wolk C (1988) Conjugal transfer of DNA to cyanobacteria. Methods Enzymol 167: 747-754

Fitzgerald N, English R, Lampel J, TJ VB (1998) Sonication-dependent electroporation of the erythromycin-producing bacterium *Saccharopolyspora erythraea*. Appl Environ Microbiol 64: 1580-1583

Halloway B (1993) Genetics for all bacteria. Annu Rev Microbiol 47: 659-684

Kahn K, Mazel D, Houmard J, Tandeau de Marsac N, Schaefer M (1997) A role for *cpeYZ* in cyanobacterial phycoerythrin biosynthesis. J Bacteriol 179: 998-1006

Moser D, Zarka D, Hedman C, Kallas T (1995) Plasmid and chromosomal DNA recovery by electroextraction of cyanobacteria. FEMS Microbiol Lett 128: 307-313

Moser D, Zarka D, Kallas T (1993) Characterization of a Restriction Barrier and Electrotransformation of the Cyanobacterium *Nostoc* PCC-7121. Arch Microbiol 160: 229-237

Mühlenhoff U, Chauvat F (1996) Gene transfer and manipulation in the thermophilic cyanobacterium *Synechococcus elongatus*. Mol Gen Genet 252: 93-100

Plude J, Parker D, Schommer O, Timmerman R, Hagstrom S, Joers J, Hnasko R (1991) Chemical characterization of polysaccharide from the slime layer of the cyanobacterium *Microcystis flos-aquae* C3-40. Appl Environ Microbiol 57: 1696-1700

Porter R (1986) Transformation in cyanobacteria CRC Crit Revs Microbiol, vol 13, pp 111-132

Rippka R, Deruelles J, Waterbury J, Herdman M, Stanier R (1979) Generic assignments, strain histories and properties of pure cultures of cyanobacteria. J Gen Microbiol 111: 1-61

Sambrook J, Fritsch E, Maniatis T (1989) Molecular clonining: a laboratory manual. Cold Spring Harbor Laboratory, Cold Spring Harbor, N.Y.

Shestakov S, Reaston J (1987) Gene transfer and host-vector systems of cyanobacteria. Oxford Surveys of Plant Molecular and Cell Biology 4: 137-166

Shimogawara K, Fujiwara S, Grossman A, Usuda H (1998) High-efficiency transformation of *Chlamydomonas reinhardtii* by electroporation. Genetics 148: 1821-1828

Sidler W (1994) Phycobilisome and phycobiliprotein structures. In: Bryant D (ed) The Molecular Biology of Cyanobacteria. Kluwer Academic Publishers, Dordrecht, pp 139-216

Swanson R, Zhou J, Leary J, Williams T, de Lorimier R, Bryant D, Glazer A (1992) Characterization of phycocyanin produced by *cpcE* and *cpcF* mutants and identification of an intergenic suppressor of the defect in bilin attachment. J Biol Chem 267: 16146-16154

Thiel T (1994) Genetic analysis of cyanobacteria. In: Bryant D (ed) The Molecular Biology of Cyanobacteria. Kluwer Academic Publishers, Dordrecht, pp 581-611

Thiel T, Poo H (1989) Transformation of a filamentous cyanobacterium by electroporation. J Bacteriol 171: 5743-5746

Wilmotte A (1994) Molecular evolution and taxonomy of the cyanobacteria. In: Bryant D (ed) The Molecular Biology of Cyanobacteria. Kluwer Academic Publishers, Dordrecht, pp 1-25

Wolk C, Cai Y, Cardemil L, Flores E, Hohn B, Murry M, Schmetterer G, Schrautemeier B, Wilson R (1988) Isolation and complementation of mutants of Anabaena sp. strain PCC 7120 unable to grow aerobically on dinitrogen. J Bacteriol 170: 1239-1244

Electroporation of Plasmids into Freshwater and Marine Caulobacters

JOHN SMIT, JOHN F. NOMELLINI and WADE H. BINGLE

Introduction

Caulobacter sp.are gram-negative bacteria common in freshwater and marine environments. These organisms exhibit a life cycle alternating between a monoflagellated swarmer cell and nonmotile stalked cell. The stalked cell is typically found attached to inert surfaces by an attachment structure at the base of the stalk, the holdfast. It was observed more than forty years ago that the spatial positioning and temporal regulation of the structures associated with this developmental progression (e.g., the stalk, the flagellum) during the cell cycle is remarkably reproducible. As a result, one freshwater caulobacter species, *C. crescentus*, emerged as a bacterial model system for cell differentiation and temporal control of gene expression (Brun et al, 1994; Gober and Marques, 1995).

Recently, *C. crescentus* has been successfully exploited for the synthesis, secretion and surface display of foreign proteins (Bingle et al., 1997a). The success of this endeavor depends on the manipulation of the bacterium's paracrystalline protein surface (S)-layer which is anchored to the outer membrane completely enveloping the bacterium. The S-layer protein is secreted by a

✉ John Smit, University of British Columbia, Department of Microbiology and Immunology, 300-6174 University Blvd, Vancouver, B. C., V6T 1Z3, Canada (*phone* 001-604-822-4417;
fax 001-604-822-6041; *e-mail* jsmit@unixg.ubc.ca)
John F. Nomellini, University of British Columbia, Department of Microbiology and Immunology, 300-6174 University Blvd, Vancouver, B. C., V6T 1Z3, Canada
Wade H. Bingle, University of British Columbia, Department of Microbiology and Immunology, 300-6174 University Blvd, Vancouver, B. C., V6T 1Z3, Canada

type I secretion system relying on an uncleaved C-terminal secretion signal (Bingle et al., 1997b). Foreign proteins linked to this secretion signal are secreted into the growth medium by the S-layer protein secretion apparatus where they can be easily recovered in appreciable quantities. Further, hybrid proteins can be engineered so that they retain their ability to anchor to the outer membrane allowing C. crescentus to be used for the display of foreign amino acid sequences on the cell surface.

Both the study of the genetics of the C. crescentus life cycle and the exploitation of the bacterium for secretion/surface display of foreign proteins depends on the ability to reintroduce plasmid-borne cloned genes into mutant strains. Early on, conjugation was the method used for the introduction of plasmids into C. crescentus, since vectors derived from most of the so-called broad host-range plasmid complementation groups could be readily introduced and maintained in C. crescentus by standard methods (Anast and Smit, 1988; Ely, 1991). But conjugation is a relatively awkward method, requiring conjugation-proficient donor strains and counter-selection procedures (often the use of several antibiotics or viruses directed against donor cells) to prevent contamination of Caulobacter recipient by E. coli donor cells. Also, the time for the complete process leading to an isolated plasmid-containing strain is lengthy, typically 7-10 days.

After Dower et al (1988) reported electrotransformation of E. coli, a number of workers successfully adapted the method for freshwater C. crescentus (Bingle and Smit, 1990; Ellgaard et al., 1989.[see Ely, 1991], Gilchrist and Smit, 1991; Thanabalu et al., 1992). Workable electroporation efficiencies (10^4-10^8 transformants/µg plasmid DNA) were obtained using various host strains and broad host range plasmids of the IncQ and IncP compatibility groups.

Gilchrist and Smit (1991) also outlined methods for the electrotransformation of marine Caulobacter sp. Many C. crescentus have an absolute requirement for salt which is typical of many or most marine bacteria. Electroporation of marine bacteria is problematic because the method requires a virtual absence of salt ions, while salt is generally required in the growth medium, and lysis of bacteria usually occurs when cells are exposed to low salt solutions.

The information presented in this report has been drawn from our published work as well as continuing experience with the organism in our laboratory and the laboratories of others in the *C. crescentus* research field.

Materials

- Electroporation apparatus (e.g., BioRad Gene Pulser™ unit with Pulse Controller™)   **Equipment**
- Electroporation cuvettes with 0.2 cm interelectrodal gaps (These can be purchased from BioRad or other suppliers, e.g., Bio/Can Scientific, Mississauga, Ontario)

- Freshwater *Caulobacter* strains (MacRae and Smit, 1991; Stahl et al., 1992)   **Bacterial strains**
    - ATCC strains: *C. crescentus* CB2 renamed J2000 (ATCC 15252); *C. crescentus* CB15A renamed JS4000 (ATCC 19089)
    - Other: See MacRae and Smit (1991)
- Marine *Caulobacter* strains (Anast and Smit, 1988; Stahl et al., 1992)
    - *Caulobacter* sp.MCS6 and MCS24

- PYE medium (for freshwater strains)   **Media**

Component	Concentration (g/L)
Peptone	2
Yeast extract	1
$MgSO_4 \cdot 7H_2O$	0.2
$CaCl_2 \cdot 2H_2O$	0.1

- Use distilled or deionized water as occasionally problems, such as poor growth rate, arise with less pure sources. Peptone and yeast extract purchased from Difco (Detroit, Michigan) or BDH (Darmstadt, Germany) is satisfactory.
- All medium components can be mixed together before sterilization by autoclaving.
- Agar is added to 1.2%-1.5% for solid medium.
- It may be necessary to supplement medium with riboflavin (2 µg/mL) for growth of non-ATCC freshwater strains.

– SSPYE medium (for marine strains)

Stock A	Component
Concentration (g/100 mL)	Peptone
10	Yeast extract
5	Stock B
Sea salts	3

Stocks A and B are mixed in a ratio of 1:50 after separate sterilization and cooling.

Note: Autoclaving sea salts (Sigma Chemical Co. St. Louis, Missouri) together with peptone and yeast extract causes a persistent precipitate. For this reason, liquid medium is made as two separate stocks which are mixed after autoclaving and cooling. Separation of ingredients prior to autoclaving is not necessary for solid SSPYE medium.

Solutions
– Distilled or deionized water, prechilled to 0-4 °C (freshwater strains)
– 10 mM $MgCl_2$/5 mM $CaCl_2$ in distilled water, prechilled to 0-4 °C (marine strains)
– 10% (w/v or v/v) glycerol, prechilled to 0-4 °C

Procedure

Preparation of plasmid DNA
1. Prepare crude plasmid DNA from E. coli using the alkaline lysis (Ish-Horowicz and Burke, 1981) or the boiling (Holmes and Quigley, 1981) methods; it is not necessary to further purify plasmid DNA.

Growth of Caulobacter
2. Inoculate a starter culture into a small test tube, typically a 16 x 150 mm size.

3. Incubate the starter culture on a tube roller until an optical density at 600 nm (OD_{600}) of between 0.5-1.0 is obtained.

4. Use a portion of the starter culture to inoculate a larger volume of growth medium in an Erlenmeyer flask (20-25% culture volume to flask volume).

5. Allow the culture to grow for at least 7-10 generations on a rotary shaker operating at 150-250 rpm (an overnight growth period is usually convenient). Freshwater strains are cultivated at 30 °C (typical generations times are 2-3 h); marine strains are cultivated at 25 °C or 30 °C (typical generation times are 3-4 h).

6. When the OD_{600} of the culture reaches 0.5 to 1.0, harvest the culture by centrifugation (10,000 x g, 4 °C).

Preparing electrocompetent Caulobacter

7. Suspend the cell pellet in one culture volume of ice-cold distilled water (for freshwater strains) or 10 mM $MgCl_2$/5 mM $CaCl_2$ (for marine strains) and follow by centrifugation (10,000 x g, 4 °C).

8. Resuspend the cell pellet in 1/2 culture volume of ice-cold distilled water (freshwater) or 10 mM $MgCl_2$/5 mM $CaCl_2$ (marine) and centrifuge (10,000 x g, 4 °C).

9. Resuspend the cell pellet in 1/20 culture volume of ice-cold 10% glycerol and centrifuge (10,000 x g, 4 °C).

10. Resuspend the cell pellet in 1/4000 of the original culture volume of ice-cold 10% glycerol. This results in a cell suspension containing approximately 10^{11} cells/mL.

11. The cells can now be used for electroporation or dispensed as 50 µl portions (in microcentrifuge tubes) and stored at -70 °C for later use. The cells can be placed directly in a -70 °C freezer; rapid freezing by use of a dry ice-ethanol bath is not necessary.

12. Add 1-10 µl of crude plasmid DNA preparation (containing 10 ng-1 µg or DNA) to 50 µl of freshly prepared cells (or just-thawed stored cells), transfer the mixture to a prechilled electroporation cuvette and electroporate using a field strength of 12.5 kV/cm. We use a BioRad Gene Pulser unit with a Pulse Controller set to 2.5 kV, 25 µFD and 200 Ω for freshwater strains or 600 Ω for marine strains. (Other laboratories report the use of 1.5 kV and 400 Ω for *C. crescentus* strains)

Electroporation with plasmid DNA

13. Immediately add 1 mL of growth medium to the electroporation cuvette, transfer the cell suspension to a 16 x 150 mm test tube and place on a tube roller at 30 °C for 2 h.

14. For freshwater strains, dilute the mixture 1/100 in PYE and plate 10-100 µL on solid medium to recover electrotransformants; isolated colonies typically appear in 2-3 days at 30 °C. For marine strains, plate 100 µL-1 mL of the undiluted mixture on solid medium to recover electrotransformants; isolated colonies typically appear in 3-4 days at 30 °C.

Troubleshooting

- The time constant is lower than expected:
  - Minimum time constants expected for freshwater and marine strains of *Caulobacter* sp. in our experience are 3.6 ms (at 200 Ω) and 10.8-12 ms (at 600 Ω), respectively. Lower time constants result in few or no electrotransformants.
  - Reasons for low time constants: electroporation cuvettes may be damaged or cell and/or DNA preparations possess excessive conductivity, typically because of excess salts from media or plasmid preparation solutions.

- The time constant is satisfactory but few or no electrotransformants are recovered when the recommended procedures are followed:
  - Occasionally *Caulobacter* fails to form a homogenous pellet upon centrifugation; the reason for this phenomenon is not clear. The pellet forms two regions, an adherent tightpacked region underlying a diffuse, loosely packed region. When the supernatant fluid is removed, considerable losses of bacteria occur during the distilled water/10% glycerol washing steps. This should be taken into account when calculating the final volume of 10% glycerol in which to resuspend the cells.
  - There is evidence that *Caulobacter* strains possessing an S-layer are electrotransformed 10 times less efficiently than strains lacking an S-layer. Penetration of the cell envelope by plasmid DNA can be enhanced by disrupting S-layer structure using EDTA prior to electroporation (see Gilchrist and Smit, 1991).
  - The plasmid used may not be able to replicate in *Caulobacter* (See Anast and Smit, 1988; Ely, 1991). This is prob-

ably only relevant to broad host range plasmids from complementation groups other than IncP and IncQ.

- In the event that electrotransformants fail to be recovered, varying the electrotransformation conditions (e.g., field strength and capacitance settings) will likely be an unproductive course of action. Gilchrist and Smit (1991) conducted a detailed study of the effect of these variables on electrotransformation of *Caulobacter* sp.; the conditions outlined above are considered optimal.

Comments

- The procedure outlined above works well for electrotransformation of *Caulobacter* with IncQ or IncP compatibility group broad host range plasmids with sizes of between 10-20 kb.

- Typical antibiotic concentrations used for selection of electrotransformants: kanamycin, 25-50 µg/mL; streptomycin, 25-50 µg/mL; chloramphenicol, 1-2 µg/mL; tetracycline, 5-10 µg/mL. Nearly all freshwater strains are naturally resistant to high concentrations of ampicillin or comparable penicillin derivatives (MacRae and Smit, 1991).

- Anticipated electroporation efficiencies: With strains of *C. crescentus*, using plasmid DNA derived from alkaline lysis mini-prep procedures, we routinely achieve efficiencies in the range of 10^6 transformants/µg of DNA. With marine Caulobacter strains, however, efficiencies are much lower and 10^3 transformants/µg of plasmid DNA must be considered the norm.

- For most applications, it is not necessary to subject freshwater strains to the extensive washing regime outlined above. The cells need only be washed once in 1/10 volumes of distilled water and once in 1/20 volumes of 10% glycerol before the final resuspension in 10% glycerol.

- We have had only limited success introducing cosmids carrying chromosomal DNA inserts into *Caulobacter* by electroporation. When it is necessary to have sizable numbers of electroporants, such as when screening a cosmid library in complementation studies, we continue to rely on conjugation to transfer such plasmids from *E. coli* to *Caulobacter* (See Ely, 1991). While this is likely due to the large size of such plasmids (ca. 50 kb), no systematic study of the effect of plasmid size or topology has been conducted for electro-transformation of *Caulobacter*.

- Cuvette re-use. Because of the high cost of cuvettes and the high usage rate typical of laboratories engaged in genetics oriented microbial research, many groups re-use cuvettes and procedures vary with respect to methods for cleaning and sterilizing. Our preferred procedure is as follows: Immediately after use cuvettes are filled with 0.1% sodium dodecyl sulfate (SDS) and extensively rinsed with water when convenient. Cuvettes are then submerged in a water-filled beaker, covered with foil and autoclaved. (Autoclaving without water seems to hasten the deterioration of the plastic portion of the cuvette). Sterilized cuvettes can remain in the water-filled beakers until they are needed, when they are rinsed with 70% ethanol, air dried and used. Cuvettes pre-pared in this manner can be reused 5-15 times. If they leak when filled with SDS solution they are discarded. The most reliable cuvettes in our experience have been those pro-duced by EquiBio.
  Although contamination from DNA used in previous experi-ments is a possibility when reusing cuvettes prepared in the above manner, our laboratory has never experienced this problem, possibly due to the fact that we do not let the cuv-ettes dry out in the presence of plasmid DNA.

▨ Applications

Although electroporation is routinely used with *Caulobacter* to introduce broad host range plasmids contained in crude plasmid samples prepared from *E. coli*, the high achievable electrotrans-formation efficiencies (up to 10^8 transformants/µg plasmid DNA) make possible a number of other applications of this tech-nique. Electroporation has been used for the following purposes in freshwater-ATCC *Caulobacter* strains:

- *Direct introduction of a plasmid from a ligation reaction*: This approach bypasses an intermediate "*E. coli* step". When *Cau-*

lobacter is electrotransformed with a ligation mix, before electroporation the ligation mix is diluted 4-fold with water, to minimize the effects of the salts in the ligation mix. The number of colonies retrieved is adequate for most purposes, but can be quite low.

- *Transposon mutagenesis*: The transposon is introduced as part of a narrow host range vector unable to replicate in *Caulobacter*. Plasmids with a colE1 replicon (e.g., "pUC-type" plasmids) are most commonly used; obviously many common cloning vectors are immediately suitable for the procedure. This procedure is done routinely in our laboratory.

- *Gene replacement/suicide mutagenesis*: A gene homologous to one on the *C. crescentus* chromosome and carried by narrow host range vector is electroporated into *Caulobacter*. Subsequent recombination can lead to the integration of the entire plasmid into the chromosome or, in the case of double crossover events, replacement of resident copy of the gene by the plasmid borne copy and loss of the plasmid vector.

- *Plasmid curing*: Some of the broad host-range plasmids can be very stably resident in bacteria and Caulobacters are not an exception. This stability is generally a useful feature but can be problematic when it becomes necessary to "cure" a strain of a particular plasmid, a procedure often necessary during mutant complementation studies. We have found that the electroporation procedure can often enhance the number of cells that have lost plasmid. In some cases, where spontaneous plasmid loss was less than 1% even after multiple sequential serial dilution cultures the simple step of using electroporation conditions increased the frequency of plasmid loss to more than 40%. Results seem to vary from strain to strain as well as for the exact plasmid construct examined.

- *Direct transfer of plasmids between bacteria*: Plasmids can be transferred directly from *E. coli* to *Caulobacter*, by mixing electrocompetent cells (*E.coli*: *Caulobacter* ratios of 1:1 to 1:5) and subjecting them to electroporation. In principle, the reverse is also possible; selection for plasmid transfer from *Caulobacter* to *E. coli* is straightforward since *Caulobacter* will not grow in Luria-Bertani (LB) medium used to cultivate *E. coli*, due to the salt concentration.

Acknowledgements. We acknowledge the contribution of Angus Gilchrist who conducted detailed studies of the electrotransformation of *C. crescentus*, Peter Awram who contributed data for plasmid curing studies and Jeff Skerker in the Lucille Shapiro laboratory at Stanford University, for providing their laboratory protocols. This research was supported by grants from the Natural Sciences and Engineering Research Council of Canada to J.S.

References

Anast N, Smit J (1988) Isolation and characterization of marine Caulobacters and assessment of their potential for genetic experimentation. Appl. Environ. Microbiol. 54: 809-817

Bingle WH, Smit J (1990) High-level plasmid expression vectors for *Caulobacter crescentus* incorporating the transcription and transcription-translation initiation regions of the paracrystalline surface layer protein gene. Plasmid 24: 143-148.

Bingle WH, Nomellini JF, Smit J (1997a) Cell surface display of a *Pseudomonas aeruginosa* strain K pilin peptide within the paracrystalline S-layer of *Caulobacter crescentus*. Molec. Microbiol. 26: 277-288.

Bingle WH, Nomellini JF, Smit J. (1997b) Linker mutagenesis of the *Caulobacter crescentus* S-layer protein: Toward a definition of an N-terminal anchoring region and a C-terminal secretion signal and the potential for heterologous protein secretion. J. Bacteriol. 179: 601-611.

Brun YV, Marczynski G, Shapiro L. (1994) The expression of asymmetry during *Caulobacter* cell differentiation. Ann. Rev. Biochem. 63: 419-450.

Dower WJ, Miller JF, Ragsdale CW (1988) High efficiency transformation of *E. coli* by high voltage electroporation. Nucl. Acids. Res. 16: 6127-6145.

Ely BE (1991) Genetics of *Caulobacter crescentus*. Methods Enzymol. 204: 372-384.

Gilchrist A, Smit J (1991) Transformation of freshwater and marine caulobacters by electroporation. J. Bacteriol. 173: 921-925.

Gober JW, Marques MV (1995) Regulation of cellular differentiation in *Caulobacter crescentus*. Microbiol. Rev. 59: 31-47.

Holmes DS, Quigley M (1981) A rapid boiling method for the preparation of bacterial plasmids. Anal. Biochem. 114: 193-197.

Ish-Horowicz D, Burke JF (1981) Rapid and efficient cosmid cloning. Nucl. Acids Res: 9: 2989-2998.

MacRae JD, Smit J (1991) Characterization of caulobacters isolated from wastewater treatment systems. Appl. Environ. Microbiol. 57: 751-758.

Stahl DA, Key R, Flesher B, Smit J (1992) The phylogeny of marine and freshwater caulobacters reflects their habitat. J. Bacteriol. 174: 2193-2198.

Thanabalu T, Hindley J, Brenner S, Oei C, Berry C (1992) Expression of the mosquitocidal toxins of *Bacillus sphaericus* and *Bacillus thuringiensis* subsp. *israelensis* by recombinant *Caulobacter crescentus*, a vehicle for biological control of insect aquatic larvae. Appl. Environ. Microbiol. 58:905-910.

Appendices

Plasmid Preparation

Natalie Eynard and Justin Teissié

Plasmids to be used for electrotransformation are extracted from bacteria by classical methods. This can be obtained by using *E.coli* as a host cell. Restriction can therefore be a problem. In such a case, there is a need to prepare the plasmids in a restriction-negative mutant of the same specie as the target cell. It is then essential to first passage the recombinant DNA prepared in *E. coli* in one of the mutant strains. Plasmids are repurified.

A key feature of a high yield in Electrotransformation is the need to use plasmids as pure as possible. Contaminating proteins and nucleic acids have been shown to be detrimental. 2 methods have been proved to be highly successful for the preparation of plasmids meeting these criteria:

- the double banding in CsCl approach

- the column method

The first approach is routinely used in all laboratories. As far as the second method is concerned, several suppliers provide all the chemical kits which are needed.

As the second one is less time consuming, it is becoming more and more popular. So called minipreparations are known to generate a significantly lower number of transformants. The purity and the state of the plasmid preparation are checked by electrophoresis on an agarose gel , staining with ethidium bromide. This method is more reliable than the 260/280 nm absorbance ratio. The plasmid must be closed circular to give high yield in transformation while linearized forms are practically ineffective in most systems.

Plasmid concentration is evaluated either by measuring the optical absorbance at 260 nm or by electrophoresis on an agarose gel, staining with ethidium bromide and comparison with the signals associated with reference amounts of DNA. After purification, plasmids are then stored frozen in TE (TE 10mM Tris HCl (pH 7.5-8.5), 1 mM EDTA).

When large amounts of plasmids are added to the bacterial suspension prior to their electropulsation, they may bring a rather large amount of ions and increase the conductivity of the suspension. This may be a problem and give some arcings which are damaging to the cell preparation and to the equipment. If this is the case, the DNA solution must be made in water.

Another problem is present when chemical modifications of the plasmids are needed. This is the case when ligations are operated. Very careful procedures are then needed to get rid of all the ions which remained after the reactions. This can be obtained by:

- dialysing the mix for 45 min (Millipore, type VS, 25 nm pore) against 1 mM EDTA/10% glycerol, pH 8

- or precipitating the DNA with n butanol and resuspending the pellet in distilled water.

Suppliers

Most products and equipments can be purchased worldwide. It is therefore impossible to give the references of all local representatives, a list, which may be outdated before this book is published. Most information is now available on the Web. Only the name of the company and its web site or Email address are given in the annex.

Only names of equipment, which was used in our groups are given. But more information can be found at http://guide.nature.com/

Anaerobic box

- Forma: forma.com
- plas labs: plas-labs.com
- B. Braun: bbraun.com
- ESI Flufrance: jouan.com
- NuAire: nuaire.com

Centrifuge

- Beckman: beckman.com
- Sorvall: sorvall.com
- Sanyo Gallenkamp: sanyo.club.tip.nl
- Jouan: jouan.com

Culture collection

- American type culture collection: atcc.org
- Belgian coordinated collections of microorganisms: belspe.be/bccm
- Salmonella Genetic Stock centre: Kesander@acs.ucalgary.ca
- Stratagene: stratagene.com
- Gibco BRL: lifetech.com
- Biorad: biorad.com
- Clonetech: clontech.com
- Invitrogen: invitrogen.com
- Specialized collections are hold in Universities

Electropulsators

- Biorad: biorad.com
- Eppendorff: eppendof.com
- BTX: genetronics.com
- Equibio
- Invitrogen: invitrogen.com

Filters for liquid sterilization

- Millipore: millipore.com
- Sartorius: sartorius.co.uk
- Whatman: whatman.com
- Gelman: gelman.com

Glassware

- Bellco glass: bellcoglass.com
- Wheaton glass: wheatonsci.com
- Sigma: sigma.com
- Bibby sterilin: bibby-sterilin.com
- Nunc: nunc.nalgenunc.com
- Corning Costar: corning.com
- Brand: brand.de

Incubator

- Sanyo Gallenkamp: sanyo.club.tip.nl
- Ibs integra Biosciences: integra-biosciences.com
- New brunswick scientific: nbsc.com
- Jouan: jouan.com
- Forma: forma.com
- B. Braun Biotech: bbraun.com
- Infors: infors.ch
- Ika: labworld-online.com
- Heraeus: heraeus-instruments

Liquid handling

- Gilson: gilson.com
- Eppendorf: eppendof.com
- Labsystems: labsystems.com
- Socorex: socorex.ch
- Biohit Oy: biohit.com

Plasmid purification

- Quiagen: quiagen.com
- Sigma: sigma.com
- Promega: promega.com
- Pharmacia: apbiotech.com

Products for culture of bacteria

- Difco: lifetech.com
- Oxoid: oxoid.co.uk
- Merck: merck.de
- Biomerieux: biomerieux.fr
- Sigma: sigma.com
- Promega: promega.com
- Stratagene: stratagene.com
- Gibco BRL: lifetech.com

Subject Index

A

Acetobacter xylinum 104, 105
- materials 106
- outline 105
- procedure 106, 107
- results 107
Actinobacillus actinomycetemcomitans, electrotransformation 182
- materials 183
- procedure 183-185
- results 185-187
Agrobacterium tumefaciens/ rhizogenes 227
albumin, serum 213
anaerobic glove box 143
anaerobiosis 141
arcing 60, 83, 133, 173, 180
ATCC 31461 108
Azospirillum brasilense, electroporation
- materials 253
- procedure 254, 255
- results 256

B

Bacillus (B.)
- B. *amyloliquefaciens, production host for industrial enzymes*
- - materials 119, 120
- - procedure 120, 121
- B. *lentus* 122
- B. *psychrophilus* 122
- B. *sphaericus* 122
- B. *subtilis, electrotransformation* 42-45
- B. *thuringensis/cereus, electroporation* 242

- - materials 243
- - procedure 244-246
- - results 246, 247
Bacteroides distasonis/uniformis/ vulgaris 140-148
BG-11 (ATCC 616) medium 260
BHA-1 medium 176
BHI-1 medium 176, 245, 247
Bifidobacterium, electroporation 72
- materials 73
- procedure 74
- results 74-76
biosafety, mycobacteria 169
biosynthesis 104
Bordetella, electrotransformation 150
- material 151, 152
- procedure 152-154
- results 154, 155
broad-host-range plasmid 88, 89, 105, 257, 268
BYGT-medium 138

C

capacitor 27
- capacitor discharge generator 69
Caulobacter, electroporation 271-273
- materials 273, 274
- procedure 274-276
cell freezing 56
cell wall 4
cellulose 104
cellulose microfibrils 105
cellulose synthase 104

Chlamydomonas reinhardtii 268
chromatic adaption 235
Clavibacter michiganensis,
   electrotransformation 221 – 225
Clostridium (C.)
– C. *beijerinckii, electroporation-*
   induced transformation 53, 54
– C. *perfringens, electroporation-*
   induced transformation 50 – 52
– DNA uptake 49
competence 107
complementation of mutants 235
conductivity 24, 284
conjugal mobilization 147
conjugation 105
culture media 36, 37
cuvette, electroporation 258, 263
Cyanobacteria, electrotransforma-
   tion/electroextraction 257, 258
– material 258 – 261
– outline 258
– procedure 262 – 265
– results 265 – 268
cyclic diguanylic activator 105

D
decay time 25
DNA, genomic, electroextraction
   264
DNA-mixture, preparation 39

E
electric conditions 88 – 92
electrical parameters 69
electrocompetent cells 8, 9
– growth medium 8
– growth temperature 8
– preparation 38, 39, 68, 80, 81
electrodes 29, 30
electroextraction 257 – 268
electropermeabilisation 11
electroporation/electroporator 50,
   258
electropulsation
– exponential generator 40
– square wave pulse 40, 69
electropulsators 30 – 32
– *Bio-Rad Gene Pulser* 73, 262
– *Gene Pulser* 119

electrotransformation parameters
   56
endonuclease 266
Enterococcus faecalis,
   electrotransformation 134, 135
– materials 135
– procedure 135 – 137
– results 137
Escherichia coli, electrotrans-
   formation 89, 105, 247
– materials 35 – 37
– procedure 38 – 41
expression phase, duration 57, 63,
   64

F
Francisella tularensis, medical/
   veterinary applications 188
– materials 189, 190
– procedure 191
– results 191, 192
freeze-thaw-cycles 138
Fremyella diplosiphon, transforma-
   tion 235 – 240

G
gellan gum 108, 116
gene libraries 231
gene replacement/
   suicide mutagenensis 279
genetic exchange 105 – 107
glucose 104
glycerol 68
– glycerol in water 43
glycine 135
– concentration 56, 62
gram-positive bacteria 199
– transformation 221 – 225
growth medium 222
growth stage 52

H
heat shock technique 109
heating problems 27 – 29

I
IM broth 73
introduction, direct 278

L

Lactococcus lactis, electrotrans-
formation 56
- materials 57–59
- outline 57
- procedure 59–61
- results 61–65
LB (*Luria Bertani*) *medium* 36, 43,
66, 110, 120, 128, 261
Legionella species, electroporation
203
- materials 204–206
- outline 204
- procedure 206–209
- results 209
ligations 284
Listeria, electrotransformation 78
- competent cells, preparation
80–84
- materials 80
- outline 79
- vectors greater than 10 kb 85, 86
lysozyme 134, 267

M

M2G medium 142
M2GSC medium 142
M10G agar 142, 143
magnesium 68
methylase 265
methylation 51
Methylobacterium extorquens,
transformation
- applications 93
- materials 88
- procedure 89, 90
- results 90–92
mutagenesis 268
Mycobacteria, slow-growing 168
- material 169
- procedure 169–172
- results 172

N

natural competence 212
Nostoc sp. 257, 262–264
nuclease 52, 265
nutrient morita broth medium
176, 177

O

osmolarity 24
oxygen 54

P

PCR (polymerase chain reaction)
258, 265
PEG (polyethylene glycol)
242–244, 248
periodontitis, juvenile/adult 182
phenotypic expression 137
Photobacterium damselae subsp.
piscicida, electrotransformation
175
- materials 176, 177
- procedure 177–179
- results 180
photo-damage 264
phycobilisome 235
plant transformation 227
plasmid curing 279
plasmid preparation 283, 284
plate counting medium,
composition 57
polyethylene glycol 134
polysaccharide, extracellular 237
Prevotella bryantii/rumicola
140–148
peridontopathogen 182
proteinase K 138
protoplasts 134
pulse duration, definition 24–27
pulse length 62
pulsing buffer (PB)
- KMR buffer 73
- LBSP 119, 120
- LBSPG 120
- PB 1 37, 67
- PB 2 37, 67
- SHMG 120
- TE buffer 37, 67

R

regeneration medium 222
restriction 216, 283
- restriction/modification system
145, 146, 247, 248
- restriction barrier 122

Ruminococcus albus,
   electroporation 195
– material 197, 198
– outline 196, 197
– procedure 198, 199
– results 199–201

S
Saccharopolyspora erythraea 267
Salmonella typhimurium,
   electrotransformation
– material 66, 67
selective medium, composition 64
shuttle vector 137, 238
SOB 36
SOC 36, 66, 67, 128, 261
somatic hybridizer 88
spark discharge 231
Sphingomonas paucimobilis,
   electrotransformation 108
– materials 109–111
– procedure 112–114
– results 114, 115
– strains and plasmids 110
Streptococcus pneumoniae,
   electrotransformation 212
– materials 213, 214
– procedure 215–217
sucrose 68
Synechococcus 257
Synechocystis 257

T
temperature 171
transduction 105
transfer, direct beween bacteria
   279
transformants 107
– expression 40
transposable elements 268
transposon mutagenesis 279
troubleshooting
– arcing 60, 83, 133, 173, 180
– frequency of transformation 84
– low percentage survival 84

V
viability 12

W
water quality 52

Y
YENB 36
YEP broth 254
Yersinia ruckeri, electrotrans-
   formation 127
– materials 128, 129
– procedure 129–131
– results 132, 133